DESIGN IS A JOB

THE NECESSARY SECOND EDITION

MIKE MONTEIRO

MULE BOOKS

ISBN: 979-8-9893587-4-8

Front cover photo: Workers of the Palace Hotel on strike in 2018 in San Francisco for higher wages, workplace security and job safety. Photo by Bastian Greshake Tzovaras. Licence: CC BY-SA 2.0. Photo has been cropped.

Editor: Lisa Maria Marquis
Proofreader: Heather Walls
Composition: Miguel Mosquito
Cover Design: Miguel Mosquito

Mule Books
601 Minnesota St. #122 SF, CA 94107
mulebooks.com

Publishing is dead. Good riddance. Do it yourself.

MORE FROM MULE BOOKS

RUINED BY DESIGN
Mike Monteiro

THE COLLECTED ANGERS
Mike Monteiro

JUST ENOUGH RESEARCH, 2024 EDITION
Erika Hall

upcoming:
THE BUSINESS MODEL IS THE GRID
Erika Hall

CONVERSATIONAL DESIGN, 2ND EDITION
Erika Hall

DESIGN IS
YOUR PASSION!

TABLE OF CONTENTS

This book is dedicated to Jeff Tidwell, for teaching me the lessons of the past...

...and to my daughter Chelsea, for giving me hope for the future.

“BETWEEN DAWN AND DUSK
THEY ARE BROKEN TO PIECES;
UNREGARDED, THEY PERISH
FOREVER.”

—Job 4:20

FOREWORD

BY SAM CABRERA

It's hard to stomach how far we've fallen since *Design Is a Job* first came out in 2012. We aren't where I thought we would be. The first edition of this book was radical—the way it spoke about design was so far from the actual practice I saw around me. In the ten years since, I've heard a lot of talk about the power of design to change the world, but without the power of the collective to act.

Frustrated, I decided to get more involved in the industry, and ran for president of the Los Angeles chapter of AIGA, the professional association for design. I was elected as (not surprisingly) the first proud, Latina woman of color to hold that post, and I am fully intent on shaking things up. Our first focus is education, shaping new standards for art in public schools in California, and getting ethics, inclusion, and the business of design into the curriculum.

If you, like me, want to be a changemaker, you probably already know that change is often met with resistance or even hostility. Support for change will come as long as the change isn't too disruptive or uncomfortable to established leaders, and those of us who push for change are seen as being divisive. But do it anyway! Your efforts will create space for others to succeed after you. You may even be asked by the asshole who inspired you to write his book foreword.

This book covers fundamentals often left out of traditional design education, but necessary for any designer to do their job well—and to truly understand their professional value. This second edition gives me hope that we will try things differently this time. Hope that we won't settle for mediocre outcomes to meet people where they are. Hope that we no longer consider it radical to want a workplace free of trauma, to guard against unethical design, and to organize for what we deserve. The second edition of *Design Is a Job* is another chance for us all to realize what is very much possible—that it's up to us to change the world.

INTRO TO THE 2ND EDITION

I THOUGHT WE'D HAVE MORE TIME.

Look, you're reading the second edition of this book. It says so right on the cover. Ten years have gone by and things ain't great. The first edition, written in the slightly more hopeful time of 2012, was about tightening our shit up. Our field was maturing, and we needed to mature along with it. We were building and building, but cracks were starting to show in the foundation. I figured I'd write a short little book to help designers do their thing, help shape this emerging new world a little bit, and then we'd all get back to work.

The cracks in the foundation were deeper than I realized. Not that people weren't trying to tell us—they were. But our own biases kept us from hearing it. I didn't realize how fucked we were then, but trust me when I tell you now: we are fucked.

I was concerned, but not even in my wildest dreams did I imagine we were working on the roads for Nazis to march back into town. We were. Here's the thing: I'm about to save some of you a lot of time. If you're getting upset three paragraphs into the intro, I'm letting you know right now that I am not relenting. You're only going to get more upset. I'm not reaching across the aisle, I'm burning down the whole damn church.

On the chance that you've been given or assigned this book by someone else, or that you picked it up from someone else's bookshelf and have little to no intention of actually reading the whole thing, I will do you the favor I've been asking my therapist to do for me for years. I will tell you the most important thing you need to know early and bluntly:

A designer needs to have a point of view.

You need to care about the shit you're working on and the effect it's having on the people and planet around you while both of those things still exist. Design has a tremendous amount of influence in the world, maybe more so at this time than at any other. Where you put your labor matters. How you impact the world matters.

If you are happy to work at a place that allows hate speech to exist, or designs databases for rounding up undocumented people, or keeps hiring the same type of cis white guy over and over and over, put down this book and stop being a designer. If you work, or aspire to work, at a place that tolerates racism, sexism, white supremacy, or transphobia, put down this book and stop being a designer. If you are afraid to rock the boat, put down this book and stop being a designer.

If you want to help make the world a better place, and believe that your actions as a designer can help make that happen, then thank you for reading this book. I hope the words between these covers can help you do these things. I'm glad you are here, and I thank you for your patience and grace.

The goal of this book remains what it was: to help you do good work. The intro to the first edition started with a very simple sentence: I love design. That remains true. I still love design. I still love designers. I hope this second edition, written in more trying times, still conveys that. And I hope I can pass some of that along to you, this time with a warning: love design, because it will love you back. Love designers, because they will love you back. Do not love where you work, because work will never love you back.

INTRO TO THE 1ST EDITION [2012]

I LOVE DESIGN.

I love it when something is well designed. I love the process of getting there. I love the interesting failures along the way. I love that I have the opportunity to earn a living designing things at a place that my partner and I built from the ground up. I love the arguing and the critiques. I love that someone I hired three years ago can now make my designs better. And I absolutely love the clients who make this possible.

Which leaves you. I love you most of all. And I am tired of seeing you get your ass kicked because no one taught you better. I am tired of you not getting paid. I am tired of you working nights and weekends. I am tired of you doing spec work because someone has convinced you it will look good in your portfolio. I am tired of you sitting by and hoping the work sells itself.

So I wrote you a book. It has a spine, and by the time you're done reading, so will you.

This book is a guide to making a living as a designer. It contains the fundamentals of being a working designer, including working with clients and others, valuing your work, and getting paid. This book will look at ways to communicate and apply these fundamentals to every aspect of your job.

The goal is to expand your view of your job as a designer to include not just your talent, but the business and communication aspects as well.

I made it a short book so you can get back to work.

This book does not contain a "system." At the end of this book, you will not run out to buy forty-three index cards. You will not get real. You will not add five items to your bucket list. You will not have to adjust your social media strategy, nor will you unlock a secret. You will have more confidence in yourself and a deeper understanding of your craft.

WHY LISTEN TO ME?

I've been running Mule Design, along with my partner Erika Hall, for almost ten years. Before that, I worked for startups, in-house marketing departments, and other studios. I've been a freelancer, a contractor, and possibly the world's worst employee. I totally lied about being qualified for the first "design job" I ever got, making logos in thirty minutes for the desktop publishing department of a local copy shop in Austin, Texas. I spent years freaking out that someone would finally figure out I had no idea what I was doing.

I started my own design studio for a few reasons. First, I was a terrible employee. I was convinced I could run a design studio better than anyone I'd ever worked for, and I wasn't sure why they had such a difficult time doing it. (Turns out the problem they all had in common was me.)

Second, I wanted control over the kind of clients I worked with. It's my belief that as a designer you are responsible for what you put into the world. When you work for someone else, you can't always pick and choose what you work on. We very consciously keep Mule small so we never feel we have to take on a job we can't ethically stand behind just to keep the lights on.

Third, I found a good partner. And I wanted to keep working with her.

Along the way we made every possible mistake we could make. We worked for free. We kept working through shifting strategy. We sat back and waited for the phone to ring. We got shafted. We lost bids for stupid reasons. We worked without contracts. I have made every mistake I am telling you not to make in this book. And I guarantee that you'll make mistakes as well—if not the same ones I did, then new and better ones. Because getting good at what you do doesn't come from not making mistakes; it comes from getting up off the floor after you fall down, and then putting up a warning sign so others avoid the pothole you fell into. This book is my warning sign to you.

I want you to do better. And work your ass off. And benefit accordingly. Because design is a job.

WHY THIS BOOK'S FOR YOU

Obviously, try as I might to be balanced, I'm coming at all of this from a bias of client services and web design. That's what I do. But whether you're in client services, or a freelancer, or working at a startup or a big company, you'll learn something from this book. Heck, you don't even have to be a designer to get something out of this. Whether you call yourself a visual designer, a graphic designer, a web designer, an interaction designer, a photographer, a casual blogger, or a modern dance enthusiast, there will be something in this book that will make you better at your craft.

"WE LIVE IN CAPITALISM, ITS POWER SEEMS INESCAPABLE—BUT THEN, SO DID THE DIVINE RIGHT OF KINGS."

—Ursula K. Le Guin

1.

WHAT IS A DESIGNER?

LET ME TELL YOU A BEAUTIFUL STORY.

You may be living it right now, you lucky dog. This story takes place in a beautiful far-away place with a deep blue sky, an ocean of green-screen possibilities, lollipop trees dripping with inspiration, and the sensuous dance of exotic muses over the techno-thump of an ambient jungle soundtrack. Yum.

In this beautiful world, there are no alarm clocks. After all, creativity demands sufficient sleep. You glide into work when you please, hopping and skipping past the insistent ogres that dare to schedule their beastly requirements meetings before 11 a.m. (there's no need to go to them after 11, either), and you slip into your Mirra chair, dismiss a few meeting requests on your wraparound 5K monitor, fire up Slack, and wait for inspiration to show itself. By then it's time for lunch, served in the cafeteria by the former chef of a Michelin restaurant your sociopathic man-child CEO ate at once and decided to buy.

In this beautiful myth, you are what is known as a "creative." You are very lucky. Because while others are weighed down by requirements, metrics, testing, and other variations of math and science, you are a child of magic (or magick, if you're a goth). Knowledge of these base matters would only defile your creative process. Your designs come from inside you.

But, of course, a story isn't really interesting until we introduce a little conflict. Shall we try a dragon? And let's say that dragon is now DMing you on Slack, excited to show you changes that were made to your Figma files, most of which you won't agree with—all of them decided in a meeting you did not attend.

The myth of the magical creative is alive and well, and it's powerful. It's equally perpetuated by designers and those who work with them. And it's destructive, reducing a designer's job to pixel-pusher, prettifier, someone who feels their way to success. A magical creative is expected to succeed based on instinct, rolling the dice every time, rather than on a methodical process that can be repeated time and time again.

Also, it makes you insufferable. Nobody likes a coworker who operates outside the rules. And this will make it harder for you to work with your team.

It also makes life harder for other designers (like me) to do their jobs. I love my job. And I don't like people who make it harder than necessary. But every time a client tells me to just "blue sky" something, or that they don't want to "stifle my creativity," I have to spend time undoing the myth of the magical creative. (Luckily, I look more like a Teamster than a magical being, but alas, you may have been cursed with attractiveness.)

Whether you are freelancing, in client services, or working in-house, you as a designer require honest feedback, real criticism, and the collaboration of other workers. That's not going to happen in a realm where colleagues or clients are worried about crushing the spirit of a magical being, or when designers don't see themselves as equal stakeholders.

So what does a designer actually do? Let's find out.

A DESIGNER IS A WORKER

You are a worker with a certain set of skills. Both parts of that sentence are important. Mostly, we enjoy focusing on the second part—the skills. Oh, the things we can do! The problems we can solve! That's the fun part, it's probably why we decided to become designers, and it's why people hire us. This book is about the other part. The worker part. The professional part. It's about how you conduct yourself. It's about how you use the skills, how you treat the people around you, and how you treat yourself.

It's also about how you see yourself fitting into this thing we call society, or community, or neighborhood, or company, or economic system. You, as a designer, as a worker, have an impact on those systems. How, when, and for whom you choose to use those skills is one of the most important aspects of your job. Those systems also have an impact on you. (The good news there is that you is sometimes a plural, and we will discuss how the plural you is stronger than the singular you throughout this book.)

Currently the economic system that most of us work within is capitalism, and while I promised my editor that I would dial it down a few notches for this book and not write a chapter on the building of and caring for your guillotine, it's fair to say that capitalism is viewed by its most ardent fans as working best when it extracts labor from its workers for as little pay as possible. Capitalism calls that profit. I call it unpaid wages. This book is here to help you get your unpaid wages, and for some of us that may be a pursuit of multigenerational accounting. And while it may be outside the purview of this book to debate the pros and cons of capitalism, it is definitely within the purview of this book to keep you from being exploited.

A designer is a worker with a certain set of skills, but for too long we let others control those skills. We let the people who hire us tell us how to do our jobs. We saw ourselves as lucky to be here. We saw ourselves as order-takers. We saw ourselves as "creatives." We saw ourselves as that most insidious of terms: individual contributors. (It's a union-busting term.)

We are workers, and we work alongside other workers.

A DESIGNER SOLVES PROBLEMS WITHIN A SET OF CONSTRAINTS

Design constraints often come in the form of material availability or unavailability (a lack of wood, a small printing press, limited bandwidth), the audience for whom the solution is intended (kids, users who aren't web-savvy, those who speak a variety of languages), and business requirements (style guides, vendor relationships already in place, our logo is Satan).

The problems to solve determine the kind of designer you are. If you're solving the problem of creating a chair that doesn't hurt your ass if you sit in it for eight hours, you're a furniture designer. If you're sixteen and holding an empty toilet paper roll in one hand and a piece of aluminum foil in the other, you're an industrial designer. If you're sitting in your office redrawing Congressional districts to ensure a better outcome for your party, you may not think of yourself as a designer, but here you are designing the end of democracy.

Since you're holding one of these smudgy newsprint zines in your hand, let's assume you're some flavor of web or UX designer, as I am. So for the remainder of this book when I pull out examples or go into specifics, they'll be from a web designer's perspective. If you're a different type of designer I'm sure you can still get something out of this, just abstract it to meet your context and apply it to your particular field.

A DESIGNER UNDERSTANDS GOALS

Whether you are helping to launch a new business or organization from scratch, making incremental changes to an existing product, or something in between, any design task you undertake must serve a goal. It's your job to find out what those goals are.

That's the first step to designing anything: ask "Why are we doing this?" If the answer isn't clear, or isn't clear to you, or just doesn't exist, you can't design anything. Stop working. Can you help set those goals? If so, do it. (Yes, it is part of your job. Anything that helps you do your job is part of your job.) How? Coming up.

A DESIGNER GATHERS INFORMATION

Who are we designing for? Are they included in the process? How will they use what we are designing? Do they need it? Can the environment support it? How might what you're designing be misused? What backend technologies does the client have at their disposal? What new ones are they open to trying? Who else has tried this and how have they succeeded or failed? Who's being included and excluded from these processes?

Once you have design goals, you need to gather as much information as possible to make sure you are designing a solution that will fulfill those goals. You simply cannot design without researching the landscape, just like you can't build a house without surveying the land.

A DESIGNER IMPOSES ORDER

Eventually all this information needs to be used to create something. Ideas and requirements become decisions. Decisions become artifacts and systems. This is the part of the job most other people will recognize as "design" because it is visible and involves pictures. And you can do it while wearing your headphones.

A DESIGNER CREATES NOVEL FORMS

Also fun. Successful design balances convention (familiar forms, terms, and interactions) with novelty—new forms to engage and delight users, in the hope they will stick around a bit longer and maybe buy their pants here instead of somewhere else. As long as you remember that those new forms must serve the goals of the business without putting additional strain on the environment. Otherwise, they're dangerous novelty.

A DESIGNER TALKS TO CLIENTS AND STAKEHOLDERS

No matter how good the work is, if you can't sell it, you haven't finished the job. I can't stress how important this is. (There's help in the appendices.) I've run into quite a few designers who left this job to someone else, be it a client rep, an art director, or a product manager. I've also seen my share of workplaces where the designer wasn't given the opportunity to sell their own work, which is amazingly shortsighted.

Selling your work directly to stakeholders is extremely important. Not only should you be able to explain why you made the decisions you made, but you'll also get firsthand feedback on where the work needs to go next. How often has your work come back to you with changes you didn't understand or agree with? And all you had was a secondhand account of what was said, or worse yet, no explanation for why those changes were requested.

Once you are ready to take the responsibility for selling your work (and I am very purposefully using the word "selling," not "presenting"), then you can begin to call yourself a designer. And get the credit for the good work you do.

A DESIGNER CARES ABOUT THE IMPACT OF THEIR WORK

At some point in the next few days you will finish this book and I want you to immediately pick up Victor Papanek's Design for the Real World, which I will bluntly summarize like so: you are responsible for the work you put in the world.

Carefully choose the projects you take on. Choose to leave the world better than you found it. Improve things for people. This doesn't mean just working on nonprofit or mission-driven projects; a lot of commercial products and services improve life for people in large and small ways. Just make sure there is some meaning to what you are doing besides exploiting a niche. Be the advocate for the person who will ultimately buy, use, or experience what you are designing.

Designers have a point of view. Your biggest value comes not from being good at using the tools, but from being good at choosing when to use them and to what end. Sometimes the best design is talking someone out of a shitty idea. We have limited resources, whether natural, financial, or cognitive. Don't contribute to people wasting them on crap.

You have more power than you think. Design not only can change the world—it has changed the world. Sadly, not always for the better.

DESIGN IS A VERB

Your toolbox should contain tools for input (what things should you make?), activity (make the things!), output (sell those things!), and impact (who benefited or got screwed by those things?). A designer's work starts way before a single pixel gets placed and ends way after the last one is locked in. You may not take the lead in every, or even any, part of the process; throughout your career you'll work on small teams, big teams, and sometimes alone. But even when you don't own a particular process, make sure to (respectfully) insert yourself. Don't wait to be asked. The more you know, the better your work will be. The better the work is, the more our communities can benefit.

Now, where do we get to use all these amazing skills? Let's find out.

DO IT FOR EXPOSURE!

2.

GETTING WORK

A FUNNY THING HAPPENED BETWEEN THE FIRST EDITION OF THIS BOOK AND THE SECOND ONE…

…the little design studios that littered the landscape got bought by big companies, and almost everyone went in-house. Which makes sense, since a lot of designers were doing a lot of work to convince nondesigners that design was important. Apparently, enough companies listened and started building up their own design teams, mainly by devouring all those little design studios.

The biggest change in the second edition of this book is reflecting this new reality. For example, this chapter used to be called "Getting Clients," and was written almost exclusively from a client-services point of view, which speaks to my own bias. Luckily, whether we are in client services, in-house, or doing some form of freelancing, the basic fundamentals of how we put ourselves out there to get work have plenty in common.

CHOOSING YOUR PATH(S)

At multiple points in your career, you will have to make decisions about how and where you want to work. Designers like to move around, and while I, personally, have spent the majority of my own career in client services, I've also done my time in-house, as a freelancer, and a contractor. I learned things along each path which have helped me along the other paths as well, so I encourage you to try them all at least once. Taking a job in-house, for instance, doesn't mean you'll work in-house forever, but it does mean you'll have experience with that type of environment—which will come in handy later if you're working at an agency and need to collaborate with an in-house team.

You might also find yourself more attracted to different paths at different times in your career depending on your circumstances. For example, the reassurance of a steady in-house salary can sound pretty sweet if you're sitting on a mountain of student loan debt. The freedom of setting your own hours might be more appealing as you get older.

But each path also comes with downsides. It's really hard to say no to the people helping you pay down those student loans, even when they're asking you to do something sketchy. Freelancing can be an anxious life of feast or famine, where you spend a considerable amount of time chasing down new work, and payment for work you've already done.

No matter where you work, you will have to weigh the merits of every project put in front of you. You'll have to decide whether you can do it well, and you'll have to understand and communicate with the people you're doing the project for. For their part, rest assured, they'll try to get more labor from you than they're paying for.

That's true even if you're in-house. On the plus side, you'll be an equal stakeholder in the organization. On the minus side, your employers will have a bigger grip on your career and, if you're reading this in the United States, your healthcare. Losing a job can have more far-reaching consequences than losing a client.

Power dynamics in an in-house situation are historically weighted toward the house side, like in a casino. But unlike in a casino, it's legal to count cards at work. Specifically, union cards. If you're choosing to work in-house, look for a place that has been unionized. If there isn't one in your area, there's a list of resources at the end of this book to help you start a union. And never work at a place that has hired union-busters.

In any case, designers need work. Which means people need to hire us, be it at an employment level or a project level. And for people to hire us, they need to find us, see what we're about, and decide to engage.

Getting work can be one of the most daunting challenges a designer faces. After all, until you've actually secured work, you can't do the job. That said, the biggest surefire way not to get work is to be anxious and insecure about not getting work. People are looking for confidence when they hire. I'm not saying you should play "hard to get." I'm telling you to behave like someone who I'd trust writing a large check to while putting my professional reputation in your hands.

Work is the lifeblood of a healthy career. It is the oxygen in your bloodstream that keeps everything going. No matter how good you are at what you do, without someone willing to pay you for that service, you will have to close your doors. Lack of work is the number one reason design careers fail. The number two reason? Who cares? (I'm kidding. It's climate change, and it's quickly climbing the charts.)

So how the hell do you get work?

PUTTING YOURSELF OUT THERE

Regardless of what path you take, at some point someone will type your name into a search box and hit return. You will be expected to be online, and to have some form of professional presence. (Your nonprofessional online presence is none of my business, and beware of working for anyone who attempts to make it theirs.)

There's no one right way to do this, but there is a right way for you to do it. The key here is to figure out how you want to be perceived, and to tailor that perception toward the kind of work you want to get. Irritatingly, you will have to figure this out for yourself.

I can, however, tell you what I might be looking for if I were the person typing your name into that search box:

- I'd be looking for a portfolio, not because I'm going to hire you based on what's in there, but because it gives me an idea of what you choose to show the world. (Pro tip: if you're a recent graduate, everyone knows you're a recent graduate. Don't try to hide it. Lean into it.)
- I'd be looking for evidence of your thoughts and writing, because this is even more revealing than your portfolio. I need to know you can write clearly. Design is about communication, and most communication is done through writing.
- I'd be looking for what kind of problems you're interested in solving, because the stuff we want to do is always a better indication of who we are than the stuff we've done.
- Finally, I'd want to know how you want to be perceived within this entity we call "design."

None of this information will be enough for me to hire you, but that's not the goal. The goal is to get me to want to have an actual conversation with you. And it's during that conversation that I might decide I want to hire you.

BE CLEAR (AND ENTHUSIASTIC) ABOUT WHAT YOU DO

Some people call this the elevator pitch. It's actually the standing-around-informally pitch. (No one wants to make small talk in an elevator. Creepy and invasive.) You need to be able to explain what you do very succinctly and in an interesting manner. Everyone hates being cornered at a party by the guy who drones on about his work.

But if you sound really excited and confident about what you do when someone inquires in a perfunctory manner at a baby shower—and then you shut up—the person you are talking to might actually remember and recommend you when the occasion arises. This actually happens all the time.

WRITE

This bears repeating: no one's going to know what you think about unless you write and publish your opinions.

I used to be incredibly shy and insecure about my writing. (Here's a secret: you don't get over it. You push through it.) Yet I knew that for the sake of my career I had to take that advice to heart. And you should too.

Write! Design! Put yourself out there. Make your opinions known so clients or potential employers can find and learn about you. People need to know who you are so they can give you money. And the more you write, the better you'll get at it. (Case in point, I'm almost certain the end of this book will be good.)

I've hired a bunch of designers myself, and while I'm always intrigued by a good portfolio (things to steal!), I've never hired anyone who couldn't show me evidence of good writing. Most of design is writing.

This does not apply to opinions on crypto or AI. Do not publish those. (We have all we need.)

NETWORK

Research is a fundamental part of design. Networking is just research plus manners. In all of human history from the time your choice of careers was either hunting or gathering, it has never been easier to figure out who is in a position to hire you, and then figure out who you know who knows that person.

Obviously, it helps to know a lot of people.

If you are not a networking natural, try a user-centered approach, just like in design. Whenever you meet someone, start by learning something about them. Try to find out just a bit about what interests and motivates them. Try to think of something you could do to help them, whether it's validating their needs and interests or providing some useful information. Then when you talk about what you do and what you need, you can put it in their context.

Good news for introverts: networking is 90% listening. Ask someone about themselves, then sit back and listen to what they have to say. Even better news: listening is the key to learning. When they're done talking, tell them what they said was very interesting. It's always true, just not always for the reason they think.

This all has to come from a place of genuine interest and confidence, or you risk sounding like a creepy stalker. As with all things, practice.

Don't be afraid to ask people you know for recommendations and referrals. Knowing someone to recommend is very exciting for people. Whether it's being thought of as karma or scoring business points, people love giving out referrals when they truly believe the person they're recommending is solid and trustworthy.

My friend Jeff Tidwell was the type of person who'd send you a postcard every few months. Sometimes it was a postcard of a place he was visiting, sometimes a collage that he'd made. Sometimes a headshot of a long-gone Hollywood star. The postcards never said much—usually a quick hello, and a short sentence to let me know he was thinking about me. I've talked to other folks who've gotten the same postcards over the years. We all cherish them. And we all know each other because Jeff introduced us. The man spent his life connecting people and maintaining relationships. He enjoyed doing it.

Obviously, I want to encourage you to maintain relationships with people you worked with—former clients, former coworkers, etc. And I want you to check in on how those old projects went, and where those old coworkers ended up. Obviously, there's a professional benefit to doing all this. But I also want to encourage you to find the joy in it. You're a good person to know! And everyone wants to introduce good people to other good people.

Maintaining relationships is not hard. (If my therapist is reading this, I just screwed myself.) You're busy people with jobs to job and lives to live, so don't reach out with chores for them to do. Send the occasional email, or—even better—the occasional postcard, just to say hello. Check in briefly. Maybe once in a blue moon make plans for coffee, tea, or a meal. And yes, make sure to tell them you're always on the lookout for opportunities, but don't make that the central point of conversation. You don't want people feeling used. Be the type of person that people think of fondly. Plus you're supporting the post office when you mail a postcard.

REFERRALS

When you need a new doctor, lawyer, or butcher, your first inclination will always be to ask someone you trust. We value the opinions of people we trust, and it's a lot faster than doing our own research.

Say you wake up tomorrow morning with a weird pain in your knee. You'll make a mental run through your contacts until you think of a friend who had knee problems last year. The next time you see her you'll ask her about the doctor who treated her, and after confirming she's not walking with a limp, you'll get that doctor's number. Now you not only have a doctor in mind, but you've gone from wondering if you can find a good doctor to hoping this particular doctor has an opening for you. All because of the wonderfully transitive properties of trust.

So it goes with hiring for design. People who hire designers are often not designers themselves. (And if this is you, please make sure to read *You're My Favorite Client.*) So in those rare times someone needs to buy design on behalf of their organization, they'll confer with friends or colleagues who've hired designers in the past. (Yes, this is also where bias may creep in). And they'll trust those referrals more than some unknown replying to a public request for proposal (RFP). In twenty years of business, most of our jobs have come through referral. Often, there is no formal RFP process. In the case of some of the largest, best projects we've worked on, we weren't bidding against anybody else.

For anyone trying to decide between two designers who seem equally capable, the one that came with a referral has a solid advantage. And that vetting goes both ways—a client or boss who is well socialized and has a good reputation in a large network is more likely to be someone you want to work for. In most cases, you're going to be as skeptical of a client who hires a designer from an ad as they are of the designer who answered that ad.

Most referrals come from colleagues or friends of colleagues, friends, or past jobs. The keys to getting those referrals are:

- Be pleasant to work with.
- Don't be afraid of conflict.
- Do good work.
- Follow up.

Get to know the people you're working with. Make them a valuable part of the project and make sure their voices get heard. People change jobs. If the current project goes well, the person who hired you will also get a reputation boost. Over time, the rest of the team will eventually spread out far and wide to other companies who will need design services at some point. Your DNA travels with them. (Not literally. I'm hoping I don't need to add a chapter explaining that.) When the call goes out to find a design partner, they'll be throwing your name in the ring.

Everything you deliver on a current project and every interaction you have with a current client is business development. All successful jobs lead to more jobs. And you are never not trying to line up your next job.

BE PLEASANT

We once received a call from a gentleman who said: "[redacted] referred me to you. He said that you wouldn't be shy about telling me I was wrong, you'd probably piss me off, and that I should listen to everything you said because it would work."

I was delighted.

That said, you should aim to be pleasant to work with, as everyone would rather work with someone pleasant than with an asshole. No one is hiring you to be their friend. They're hiring you to design solutions to problems. But if they can get the same solution from someone who's pleasant and someone who's a jerk, they'll go with the former.

DON'T AVOID CONFLICT

There's a difference between being enjoyable to work with and being "nice." Being nice means worrying about keeping up the appearance of harmony at the expense of being straightforward and fully engaged. No one wants to work with someone who's faking it. Sometimes you need to tell someone they're making the wrong call. Part of being a designer is being able to do that without coming off as a dick. But being afraid to do it because you're too invested in being "nice" is worse than being a dick.

Doing good work often requires a few hard conversations. The hard conversation you're avoiding today will turn into a much harder conversation later. Tomorrow, or somewhere down the line, another worker will end up footing the bill for the problem you're deferring on today.

DO GOOD WORK

Of course, being the most pleasant person in the world won't help your cause if the work isn't good. Just don't make the mistake of thinking the quality of your work by itself will be a shining beacon that gets you an endless parade of jobs.

Good work is the core of your career, and, for the purposes of this book, let's assume you do good work. But people aren't hiring your portfolio; they're hiring you. So while your portfolio is important as proof that you can do what you say you can (especially if you can match the work up with success metrics!), it can't be your biz dev department. You need to convince your potential clients that you'll be able to solve their problem as well as you solved your past clients' problems.

To do this design thing right we're going to have to redefine what we think of as "our work." That stuff in your portfolio? That's just evidence. The real work is that plus all the conversations, decisions, and convincing you did along the way.

REFLECT WELL ON PEOPLE WHO RECOMMEND YOU

When a friend or colleague recommends you to a potential client, you carry a responsibility to that referrer. You need to do right by them. They've put their reputation on the line to vouch for you. Treating that referral well not only shows that you're a reliable person; it also grows your referral network by one more happy client.

Likewise, sloppy work on your part reflects badly on the person who recommended you. Not only are you less likely to get another lead from that person, but you also might not be able to use them as a reference, and you've potentially added strife to the relationship between two of your clients.

Do you have to take a job that came from a referral? No. You still need to go through your process of deciding whether the job is right for you or not (more on that coming up). But how you handle the inquiry is key. When an inquiry comes in from a referral, the very least you can do is set up a ten-minute call. In ten minutes you can usually figure out whether you can or can't help this person with their problem.

Some of you might recall that the first edition of this book had a link to a screener. (Thanks for buying both editions, by the way.) It was a simple screener: a few questions like what are you trying to accomplish, why, and can you pay for it. Over the years, we added more questions; we saw other studios do their own screeners, and stole questions from one another, and that was all fine. But ultimately it led to a prospective client being given a five-page form to fill out just to talk to someone. No one wants to do homework before giving you money.

I stopped using the screener and just started emailing people back and setting up short calls. During these ten-minute calls, I can tell if the project isn't a good fit, and I can recommend someone else they can reach out to or ask if they want an intro. The referral is thrilled, and I have another human being telling everyone I was helpful. Win-win.

The most important thing to keep in mind during referral calls is that you are evaluating the potential job as much as they are evaluating you. Prospective clients sometimes find this surprising. Those folks aren't likely to make good clients.

Even when you choose not to work with someone, you want to leave them with a good impression. Someday they may have a project you'll be right for, or someone they know might, and they can refer you.

DIVERSIFY IS A VERB

Referrals aren't all wine and roses. Depending on the biases of the people getting and giving those referrals, you could be looking at a lot of people hiring people who make them "feel comfortable," mostly because they look the same, share cultural touchpoints, shop at the same stores, or went to the same school. Which explains why most companies are 80 percent white dudes named Chad. (If your name is Chad, don't get mad at me—get mad at your parents.)

So if you're in the business of giving referrals, this is a good time to take a look at the folks you keep referring and see if you can spot a pattern; patterns can be changed. If you're in the business of getting referrals, you've probably got enough privilege stored away to point out to Bill that everyone he refers to you is a white male. (Yes, it should bother you if all the referrals are white males. If it doesn't, please get out of the business of hiring people.) How Bill replies to this will also tell you if you want to continue asking him for referrals or referring him to others. (By the way, as soon as you're done reading this book I want you to pick up David Dylan Thomas' *Design For Cognitive Bias*. Don't talk to another human being outside your home until you've read the entire thing.)

OTHER, LESS GOOD, WAYS TO GET WORK

Although a lot of your work will come in through referrals, there's no reason to sit on your hands waiting for the phone to ring. There are other, if less good, things to be doing in the meantime. Let's review your options.

Requests for proposals (RFPs)

People who tell you they don't go after RFPs and people who tell you they don't have meetings are both lying to you. The problem with most RFPs is that they're written by people who are as irritated that they have to write them as you are that you have to reply to them.

That's also the key to handling an RFP. Find out who wrote it. Most come with a contact number in case you have questions. Call it. Make friends with the people who wrote it. They may have even heard of you. If so, you're golden. If they haven't heard of you, this is a good time to practice that elevator pitch. Strike up a conversation with them and get as much detail as you can about the organization involved.

The other problem with RFPs is that they can be overly prescriptive in nature. They can include specific solutions that may or may not be appropriate. This is usually an indication that the organization is freaked out about having to hire designers and is trying to maintain as much control as possible—to the point where they're going into as much detail as possible and short-circuiting the entire design process. You may be able to help them through that.

If an RFP starts dictating button colors, pick up the phone. It's a cry for help and an opportunity for you to help create mutual understanding. Speaking directly to a designer may be just what they need—much more than getting 325 replies to a badly formed RFP.

Remember, not all organizations who send out formal documents have to, but some do. There are many good organizations that have to go through an RFP process to hire anyone. And those are good jobs to go after.

By the way, how do these organizations decide where to send these RFPs? By referral. BAM!

Outbound contact

I totally encourage you to go after jobs you want. Let's just be realistic about the return on this type of business development: it is very, very low. You're basically throwing seeds across a cement lot in the hope that one or two of them will find a crevice with enough dirt to take root in, not get eaten by birds, and experience rain at some point. It's hard enough to get a job that exists, but contacting people when a job may not exist is rough.

That said, if there's a company you're really interested in working with, go after them. Your best bet, as always, is going in through your network. Someone you know will know someone at that organization. Get ready to buy some meals and drinks. This is a little like that game where you start off with a paper clip and have to trade your way up to a dream client in ten moves or less. But with a little luck you may be able to get in front of the right people. More likely than not you'll be making a pitch for them to consider should they have design needs in the future, so make a lasting impression.

Oh, and potential clients love getting cold calls as much as you do.

Advertising

Mule was once asked to cosponsor a party at SXSW. I believe it was about three or four grand to get our name on the banner outside the bar and a listing in the program. We didn't have a spare three or four grand sitting around at the time, so instead we got stickers with our logo on them. Cost: $50. (It helps that our logo is an animal. People like animals.) We went to the party and gave everyone a sticker. People put stickers on, took pictures, and uploaded them to the internet. You know how people referred to that party? The Mule party.

So yeah, I believe in advertising. I also believe in not throwing money away. Did we get any clients out of that party? Probably not, but we raised our profile a little bit. And we were only out $50. Oh, and we drank the free beer paid for by the people who anted up the $4K.

Host the occasional party, buy the occasional ad in a conference brochure. Be visible in places where clients look. Very specifically targeted advertising might be helpful to raise your profile and let people know you exist. But I wouldn't put more than one or two eggs in that basket if you're trying to get clients. At the most, advertising might help you seem familiar to a potential client who's just been referred to you.

Conferences

Conferences are a fine place to meet both potential clients and peers (sometimes they're the same person.) They're full of other designers who work at organizations big enough to have a conference-going budget. (They're also more likely to expense dinner for a whole table, so tag along.) Those are the same large organizations who hire outside designers and send out RFPs, at which point knowing someone on the inside will be invaluable.

More writing

If you're interested in working with Disney, then blog about Disneyland. Write the best essays about Disney design on the internet. Will this in and of itself guarantee that you'll be the designer they call when they need work? No, but it certainly won't hurt. And you'll be writing about what you love anyway.

Yes, this is the third time I've mentioned writing. It's not an accident.

Gig marketplaces

Gig marketplaces—by which I mean Fivvr and their ilk—are technically an option. Look, if someone is going to pay you fifty dollars to draw five hundred bored ape bullshit NFTs, they're never going to pay you $5,000 to draw anything. They're just going to move on to the next sucker willing to do it for fifty dollars. And you will always be the dope who spent their time drawing five hundred bored ape bullshit NFTs for fifty dollars.

I absolutely understand everyone's need to pay rent, so do it if you need to do it. Just know that when you devalue your labor, not only does it tend to stay devalued, but it also sets the value for that same labor from other designers.

THE INTERVIEW

At some point, all or one of these efforts will pay off, and you will find yourself sitting across a table from one or more people, trying to get a job or a project. They will have done their homework on you. You need to do your homework on them.

Before you sit down at that table, know how they make their money. If they're a startup, know how much of that money they have left. (All of this is on the internet.) Know what they do and how long they've been doing it. Know why you're sitting there. Are they hiring because they're growing, or because the last person moved on, or because they run through designers like my little dog runs through chew sticks? Ask for the names of the people you'll be interviewing with. You may not get them, but if you do, take a look at their professional online presence. Learn a little about who you're dealing with.

The most important thing to remember about an interview or pitch meeting is that you are interviewing them. You are finding out if this is a place, or a project, that you are willing to sell your labor to. They may think it's the other way around, and that's okay. You know the truth. Your labor is the thing that makes this machine go, and you need to be very careful where you put it.

So ask questions. Have more questions for them than they have for you. If a potential employer is annoyed that an employee is too inquisitive, finding that out during the interview is better than finding it out after you've completed your health insurance paperwork and stopped looking at other offers.

If you are bringing work to show, think of the work as a conversation starter. The quicker you can pivot away from your portfolio and engage the audience in a conversation about their problems, the better. It means they've decided to trust you. Lean into it. When we first started doing pitch meetings, we measured how well the pitch went by how far we got in the pitch deck. If we got all the way through it, we weren't getting the project.

The dumbest thing I ever did at an interview was tell a prospective client that their problem was easy to solve. I wasn't lying—it was, for someone with our skill set. But they'd just finished telling us how much they'd struggled with it, and being told they were struggling with something easy made them feel stupid. Never pass up an opportunity to empathize with people. You are sitting across the table from people with a pain point. The first step in taking someone's pain away is to acknowledge that it is real and it causes pain. Plus, you want them to think of you as someone who fixes hard problems, not easy ones.

Depending on where you're applying, they may ask you to perform a task. Again, I'll leave this to your discretion because you know your situation better than I do, but I think it's worth asking what they're hoping to learn from the task that they don't already see. A lot of companies give these tests because at some point someone in HR read a book about it. Challenge it. The people who complete the task get a box checked off. The person who challenges the way we do things might be our next design director.

If you're looking for a job, you should be collecting interviews like Pokémon. Do as many as possible, starting with ones you might not be too psyched about. They're good practice, and by the time you get to the ones you're really interested in, you'll be in your groove.

And for the love of all that's holy, never give your interviewer or potential client a gift, no matter what your design professors told you.

TAKE YOUR SPACE, AND GET YOUR FLOWERS

If you are going to make a career of this, you need to be seen and you need to be heard, which means you need to be willing to take up your space, and to do so for the benefit of others. Let people know who you are and what you stand for—and for fuck's sake, be willing to stand for something. As Fred Rogers said, in times of trouble, people look for the helpers. Be the person people think of when they are looking for the helpers.

As always, the key to everything, especially getting work, is confidence. No one wants to deposit a check into an ATM that looks like it just got wheeled in place yesterday and may not be there tomorrow. Ultimately you need to evaluate whether a client is the right client for you, because the perfect client is one who understands and values what you do, whose design problem plays to your strengths, and whose timeline matches your availability.

"YOU HELPED TO FUEL THE DEATH MACHINE THAT DOWN THE LINE WILL KILL YOU TOO."

—Run The Jewels

3.
CHOOSING WHO TO SELL YOUR LABOR TO

GETTING WORK IS GREAT. BUT IT TENDS TO COME IN ENOUGH DIFFERENT SHAPES AND SIZES THAT SOME OF IT WILL NOT BE RIGHT FOR YOU.

Or you may not be right for it. Or what the people needing the work want to accomplish may not be right for the planet. For the sake of everyone involved, you need to make sure you're putting your labor into the right work.

You have a responsibility to your potential client or employers to make sure you're actually the right designer for the job. You have a responsibility to yourself to make sure the work you're signing up to do is within your capabilities. And you have a responsibility to the community to ensure the work you're doing isn't detrimental to that community. If you're responsible for bringing in work for a shop full of people, or you're the in-house team lead, you certainly have a responsibility not to put your workers in a bad position by bringing in work that doesn't play to their strengths, or puts them in an ethical quandary.

The work you choose to take on defines you. Your portfolio and résumé need to tell a story, and all the work you add to them becomes a chapter in that story. Make sure you're consciously building the story you want to be telling. And make sure your story is compelling enough that your next client or employer is excited to become a character in it.

In twenty years of business at Mule, we've taken on projects we've been personally excited about, projects we thought had a fantastic benefit to the world, projects we thought were great fun, projects with subject matter we were curious about, and yes, even projects that were taken primarily to keep the lights on (if any clients are reading this, please rest assured that yours did not fit into that last category). But we've never taken on a project that we either were ashamed of or knew would be doomed from the start. In the interest of full disclosure, we've certainly taken on projects that we thought were solid, and had them turn into nightmares. No system is completely foolproof. (Again, if any former clients are reading…)

YOUR ETHICAL RESPONSIBILITY

A few years ago I was interviewing a designer for a job. We ended up passing on him. Not because the work was bad; it wasn't. And not because he interviewed badly; he didn't. We passed on him because, while reviewing his portfolio, we came across work for a client I won't name. Let's just say their product kills people.

"Why did you work on that?" I asked him.

There are two answers I would have accepted. The first would've been: "I don't have a problem with their business model—in fact, I think there are too many people in the world." Weird? Sure. But hey, he would have been taking a stand. The second answer I'd have accepted would've been, "They're terrible, but I really needed the money and had no other options." I can't begrudge anyone making a living, and we've all done a few things we're not proud of.

But instead, he looked surprised that I was asking the question, and said something to the effect of it simply being the next project on his plate.

I asked him if he agreed with how the client made their

money. He replied in the negative—he'd just done the design.

I told him we didn't take on any projects that we couldn't ethically stand behind.

And here I'll quote him: "Must be nice."

And that's when I decided not to hire him.

Now, I'm not trying to embarrass this individual, who I'm sure went on to have a solid design career; I might have even answered the same way at his age. But as a designer, hell, as any type of craftsperson, you are responsible for what you help to put in the world. You are defined by the clients you take on, and you can only stand as proud of the work as its benefit to society entitles you to.

Let's picture ourselves on the other side of this equation, because as private individuals we deal with professionals all the time. Say, for example, that you're preparing dinner for your family and stop at the fish counter of your local market. You ask the fishmonger what's good. They point at the tilapia. You ask them to wrap up three pieces, take them home, cook them up, and feed your family. Your family gets sick. You're angry at the fishmonger for selling you bad fish. I would be too.

Here's what you don't know: earlier that day, the fishmonger had tried throwing out that tilapia because it was ripe, but the market manager stopped them.

"I don't care if it's bad. Our margins suck. Sell it anyway."

"But someone might get sick."

"I'll fire you if you don't try to sell it."

The fishmonger didn't want to get fired, so they sold the tilapia. Now your family is sick. You're probably thinking the fishmonger should've stood up to the store manager, right? But at the same time, as designers we often find ourselves in that position. And how many times have we done exactly what the fishmonger did? How many times did we look at something we were tasked with designing and know it was harmful? How many times did we just design it because we were afraid of having an uncomfortable conversation? The products we design are complex and insidious enough that when we don't stand up for the health and well-being of the people who interact with our work, those people can die.

Be the fishmonger you want to see in the world. Do your work with the same amount of respect for others that you'd want them to show for you.

Before you take on a project or accept a job offer, ask yourself whether the problem you're being asked to solve is one you feel good about attaching your name to. Does it serve a real need? There is absolutely nothing wrong with making money—in fact, one of the goals of this book is to encourage you to make your fair share—but making money to someone else's detriment makes you complicit in that person's suffering. If a product you design does harm, then you have done harm.

I urge each and every one of you to seek out projects that leave the world a better place than you found it. But Mike, we need to keep the lights on! I agree, but if the only way you can keep the lights on is by taking on work that hurts other people, you're in the wrong business. Please find another.

BEWARE OF BAD CLIENTS

The business development process should go both ways. A potential client or employer is trying to decide whether they want to work with you, and you're trying to decide whether you want to work with them. Can they pay you fairly? Are they tackling a problem that's interesting to you? Do you have the core competencies to solve it? Is there room for you in the problem-solving process? Are the people experiencing the problem involved in the project?

Over the years, the one constant that Mule has been able to rely on is that potential clients behave better in the business development process than they do during the project. Trust your gut. If your potential client or employer is slow to return your calls now while they're trying to engage you, they'll be slower later. If gathering requirements or technical constraints is hard, then gathering feedback will be harder. If talking about money is challenging now, it won't get any easier once they actually owe you some. If your early conversations reveal a ton of red flags, then disengage. You will not be able to do good work, and neither you nor the client will be well served.

Optimally, a lead shows up and says, "We have a good sense of our problem, we're not tied to any particular solution, and we're looking forward to working together to come up with one using our collective expertise." And then we all take off on our rainbow unicorns.

But more often, leads tend to show up with a solution in mind, and they can't fully articulate the problem. It's whether you're able to walk them backward from their solution to the actual problem, and how open they are to that process, that lets you know they're going to be a good client. Your ability to do that is a sign of how good a designer you are as well.

That conversation generally goes something like this:

"The homepage just needs two big buttons: 'buy pants' and 'subscribe to our newsletter about pants.'"

"What's your goal with those big buttons?"

"We want people to buy our pants."

"Are they good pants?"

"They're the best pants."

"Why don't we start by getting people as excited about your pants as you are?"

"Go on…"

That's gonna be a good client.

A bad client conversation goes like this:

"The homepage just needs two big buttons: 'buy pants' and 'subscribe to our newsletter about pants.'"

"What's your goal with those big buttons?"

"What do you care? Just make my buttons!"

Bad client. Walk away and let them see how great you look in your pants as you do so.

Beware of clients or employers who have waited to hire you until they have a perfect diagram of what they need and want you to color it in. If they're not coming to you for strategy and problem-solving, then they're not coming to you for design, they're coming to you for production. And if you take on production work, you'll get known as the person people can keep going to for production work. That might pay the bills, but once you're known for that, it's hard to get people to think of you as a person you can go to for more than that.

Look for clients and employers who have clear goals, not detailed punch lists. This is especially true of RFPs that require you to reply directly to each line item at the risk of being disqualified from the process. You don't want to sign up for a process that you know is broken from the start. Once you set sail on a boat, you can't convince a captain to take to the sky.

Pick-up artists

You know that douchebag at the bar who walks up to your friends and says, "You know, I usually date models..." Yeah, that guy. The client-services version of that is, "You know, we've got some really big-name agencies who'd love to get this job." Great, go call them. Don't work for someone who tries to make you feel like they're lowering themselves to work with you, even as a negotiation tactic. Good work comes from mutual respect.

Reluctant buyers

Never work with someone who shows up begrudgingly and doesn't understand the value of design. It's not going to work out. If someone doesn't understand the value of what you do, they're going to be reluctant both to do the right work and to pay the right price.

We see this most often with early-stage startups who were sent by their VCs. The startup people don't want to be working with a design agency; either they don't see the value, or they feel like bringing in an outside resource is slowing them down. We also see it with internal designers who were sent by a higher-up—they feel like they should have been allowed to do the job themselves, or they're freaked out about their own job security. Either way, a reluctant buyer will be focused on proving that you weren't needed, constantly trying to disengage themselves from the process like a little kid tanking the violin lessons his dad forced him to take.

The design process gets messy, and people need to be able to tell each other difficult things like, "Your business model exploits your drivers, who are critical to your success and deserve a fair share of it." You won't be able to have those discussions with people who don't value what you have to say.

Opportunists

Someone who is in the business of selling pants will still be selling pants in three to four months, provided they're still selling anything. But when someone selling pants comes to you with an idea to sell toasters, beware. Why are they getting into a new business? Is it a natural growth of their existing business? Does their expertise in their current business translate to their new business, or is it an opportunistic attempt to "fill a hole in the market"? That hole may close before the project is over, and they'll want to move to the next opportunity. Want to redo all that work? No, because they'll want you to do it for free.

MAKING CLIENTS BETTER

Now, it's easy to laugh at clients and say they're bad clients. But the truth is that no one is born knowing how to be a good client, just as no one is born knowing how to be a good designer. And look how hard you have to work at being a good designer!

By and large, most clients want to be good clients, and they're trying to do the right thing by their business.

Meet them halfway

Clients will always ask you to make their logo bigger, prescribe solutions, and ask you to do things that will make you smack your forehead. You can roll your eyes at how much they don't understand about design, or you can roll up your sleeves and begin practicing your craft by helping them clarify what they need.

Not knowing the design language doesn't make someone a bad client. I doubt very much that most of you could have a medical conversation with your doctor on par with a conversation your doctor could have with another doctor, and that doesn't make you a bad patient. And just like a good doctor can put you at ease with a sensitive bedside manner and thoughtful professional terminology, a good designer needs to cultivate a productive means of getting the necessary information from their clients.

It's your job as a designer, and as a communication professional, to find the right language to communicate with your client. When you say a client doesn't "get it," you might as well be saying, "I couldn't figure out how to get my point across. I am a lazy designer. Please take all my clients from me."

Remember who you work for

Regardless of where you end up working, whether it be in-house or client services, a big company or a startup, the public sector or civil services, there is one thing that will always be true: you don't work for the people who sign your checks. You work for the people on the other side of the glass. You work for the people affected by your work. They're the ones paying the price when something goes wrong, and they're not in the room when decisions about them are being made.

Which means it falls on you to represent them and to look out for their well-being. Remember our fishmonger? Whose interests did you want them looking out for—yours or the market manager's? Your expectation was that they were looking out for you, and they should've been. You were right to be upset at the fishmonger, but only if you're willing to behave the way you wish they would.

WORK FOR MONEY

You are in business. You need to be as confident about money as you are about design.

How clients or employers talk about money is also a good indication of what they'll be like to work with. Do they already have the money? Do they have to make the business case for design to get the money? What do they value and what are they willing to pay for? What don't they want to pay for? Do they have access to as much money as it will take? Do they already have a budget limit? Have they already worked with a designer who took their money and didn't give them value in return? Are they cagey about telling you how much they want to spend? If so, why?

Beware of clients who want you to work for equity or other things that can't be easily converted into rent. There's too much there beyond your control. And especially beware of clients who tell you the work you do together will look great in your portfolio. For one, it probably won't. A client who asks you to work for less than market rates already disrespects you. How do you think that'll play out during the project? Think you're getting portfolio-level work out of that? Secondly, you're in business to make money. I'm lucky enough that I get to do what I love, but only because I can make a living from it. Working for portfolio fodder is the same as dying young and leaving a good-looking corpse.

Beware of working for free. Any work you take on for free will get pushed aside for paying work. That does neither you nor the client any favors. Neither of you will respect each other's time. If the situation merits it, work at a discounted rate. But submit a budget showing the actual rate, with the discount applied. Let the client know the value of what they're getting.

Does this mean you should never work for free? Work is an exchange of labor for capital. If the person you are laboring for is earning capital from that labor, then you are due a fair share of that capital. However, if, say, your neighborhood gets together and decides they want a localized nonracist version of NextDoor—well, that's a fine thing to donate your labor to, because you're contributing to and benefiting from a community good. But if Pat starts inundating that community good with pyramid-scheme leggings, it's time to have a conversation with Pat.

Money is a standard part of any business transaction. Watch out for clients who don't understand the financial value of what they're hiring you for. (Or attempt to convince you they don't.) In 95 percent of the jobs we've taken on, our clients have met or exceeded their success metrics after our work was done. Can we take full credit for that? Sure! Well, no, probably not, but it's safe to say we were definitely a part of it. Design, done right, is not a loss on the balance sheet and should not be bought or sold as one. Design is an investment in infrastructure and keeps the wheels of business running smoothly. Good design equals a more effective product or service. Design means workers get paid, and customers get served.

When someone tells you that design is your passion, they are about to fuck you. Design is your job. Jobs are labor. Labor gets paid.

Oh, and by the way—crypto is not money.

START WITH RELATIONSHIPS

Whether you're working internally or externally, make sure you can spot the differences between a situation that can benefit from your help, an otherwise good situation that's just not a good fit for you, and an actual bad situation.

Will following these rules help you do better work? Probably. Are they fail-safe? Oh, hell no. But they're based on many, many years of trial and error, messing up, dusting myself off, and trying again. As always, I'm not advocating for a single way of getting things done; I'm telling you how it worked out for me. Your mileage will certainly vary.

Design careers are stories based on a series of good relationships. Before you enter into one, make sure that all the necessary elements for success are there. You'll be spending some time with these people, and all your reputations are on the line. Above all, the best advice I can give you in selecting your clients is to be confident, treat everyone with the same respect you want them to treat you with, trust your gut, and do your homework.

Now that we've decided who to work with, let's figure out what to charge them.

CORPORATE PROFITS
ARE UNPAID HAGES.

4.

THIS IS THE CHAPTER ABOUT MONEY

THIS SHOULD BE THE SHORTEST CHAPTER OF THE BOOK.

Anything I have to tell you can be summed up thusly: charge as much as you can, deliver an honest value, corporate profits are unpaid wages, and (almost) never work for free.

Unfortunately, since most designers feel such pangs of guilt about charging an adequate amount for their labor, and capitalism is designed to extract the most amount of labor for the least amount of pay, it's probably worth the time and effort to go a little deeper.

Our responsibility for what we put into the world works both ways: we deserve blame when our work hurts someone, and we deserve credit when our work succeeds. We are answerable for both success and failure. In other words, if you're gonna glare at me when shit goes wrong, you better be writing extra checks when shit goes right.

I'll be up-front: most of the examples in this chapter will resonate stronger for freelancers and folks in client services, but the abstract lessons on confidence and value are applicable to you no matter where you work.

LET'S GET COMFORTABLE TALKING ABOUT MONEY

It has come to my attention that some of you are uncomfortable with the idea of money. Let's start by going over a short list of people who are not uncomfortable with money: your landlord, your utility companies, the therapist helping you get over the trauma your employer is dishing out, your insurance agent, your grocer, the assholes the government sold your student loan to, and your doctor. If you have kids, let's add their orthodontist and the bursar at their eventual college of choice to the list. Money is so necessary to the fabric of society that even my friends in Berkeley have a secret guilty stash of it.

The minute you got into the design game, you took up a trade that involves money changing hands. And hopefully you can get on the receiving end of that transaction. You are not doing design, you are selling design, which is a valuable service. If you don't want to charge for your services, you can pick up Design Is My Passion in the giftshop of the next immersive Instagram experience that rolls into town at the place where the Spirit Halloween Store used to be.

But if you are trying to make a living as a designer, then comfort with money needs to be part of your skill set. Whether you are putting together a proposal for a project or negotiating a salary, there will be no ums, no hesitation when a someone asks for a number, no asking them to tell you what your work is worth. As with all parts of the job, confidence begets confidence, and it's even more so with money. No one wants to hand money to someone who behaves like they don't deserve it or know what to do with it.

And above all, you will never, ever tell a client or an employer, "I'm not comfortable talking about money."

If I'm hiring someone to fix a problem, design a product, or build a design team, I need to know that I'm hiring a competent person. Résumés are nice, but they take time to read, and people are busy. We associate confidence with competence; competent people are confident in their abilities. And since the person hiring you is probably confident with money, they'll see your display of confidence as a display of competence. If you're asking for a lot of money, it must be because you're very good at what you do. (Consultancies like McKinsey are built on nothing more than this.) Your comfort with money is a display of competence, in a language that the people hiring you are very familiar with. Think of it as a gift: you're saving them time.

So how much should you charge? How much salary should you be making? As much as you can. The problem with designers isn't that they don't know how much to charge; it's that they're afraid to charge it.

You can always negotiate down from a price or a salary demand that was too high, but you can't recover from one that was too low. You're stuck with that. There is nothing wrong with trying to get the highest price for your work. It's supply and demand. Capitalism doesn't deserve your charity. Capitalists can't get mad when someone is playing a game by the rules they set. The more demand there is for a particular service, the more suppliers of that service can charge. You'll find that your rates will fluctuate as the market for design ebbs and flows. But if you are not confident in the value of your work, there's no way you'll be comfortable charging a fair price.

APPROACHING PRICING

A few years ago I was fortunate enough to work with Kick-Start International, a non-governmental organization (NGO) that designs and manufactures low-cost water pumps for use in impoverished agricultural areas of the world, mainly in sub-Saharan Africa. They have an amazing track record of helping people lift themselves out of poverty by using these simple, easy-to-fix water pumps to irrigate crops. They create jobs.

Here's why I'm mentioning it: they don't give away the pumps, like most NGOs would. They sell them.

The KickStart founders spent years working with NGOs who donated equipment and tools to those in need, only to return to the scene and find that equipment had been scavenged for parts or was sitting unused and rusting away. People didn't value (or need) what they had been given for free. So KickStart decided to sell their pumps, marketing them as the "Super MoneyMaker."

The results were impressive. The Super MoneyMaker became an item that area farmers would save up for. Only people who actually planned to use one would buy one. Most telling, people would scavenge from other things to repair their Super MoneyMakers. When people pay for something with their own money, they value it more than when they get it for free. They take care of it.

Undercharging for your work has the same effect. Clients value you in direct proportion to how much it costs them. If you have ever done volunteer work, you may have noticed that the people you were volunteering for had no trouble making increasing demands on your time. No one is going to value your work, or your time, if you don't.

Oh, and guess how much people value free work? You got it. Your design work is your client's Super MoneyMaker.

CHARGING FOR PROJECTS

Back when we used to make websites, we'd get emails and calls asking, "How much do you charge for a website?" And while it's easy to roll your eyes at a question like that, you'd do better to remember two things: you're in a service profession, and it's a fair question. From a client's perspective, the basis for pricing is opaque at best.

Design can mean a lot of things. Ask five designers what they would charge to design the exact same thing and you will get five different answers. To a prospective client, the idea of hiring a design firm can be daunting enough without having to figure out how much it's all going to cost. Calling a bunch of designers and getting a range of prices that don't resemble one another has got to make you feel like people are just making this shit up. Which they are.

A client can get work of any description for a cost between zero and infinity. And people who don't buy design every day might not have a good idea of all the options that go into setting a price. It's your job to help them figure it out. And sometimes the best way to do that is to help them map it to something they are familiar with purchasing. After all, there's not much difference between asking, "How much do you charge for a website?" and "How much do you charge for a car?" Except that people have probably bought a car before. They know that the type of car a college student needs to get back and forth to class is vastly different than what parents with a new set of twins need. They know that they'll have to pay more for the latter. And they know that, even among one type of car, the price can vary wildly based on their quality and feature set, which really isn't that different from decisions you'd have to make in choosing what kind of website, app, or illustration you want and who you want to build it. Except they might not know that yet (metaphors are handy).

When you first go off on your own to freelance or start a studio, you might have the idea that everyone else has it all figured out. (We totally do! I'm totally lying!) If you're applying for a job, especially at a small company or a startup that's never hired a designer before, you might be surprised when they ask you what designers get paid. There's no secret formula for knowing how much to charge for your work. There's only the experience you gain from having to do it over and over again. You'll undercharge a few; you might even overcharge a few. But eventually you do it enough times that patterns start to emerge. You gain enough experience to realize that this new job is similar to a job you did last year, which you undercharged for. So you make the correction. Eventually what you gain is the confidence to trust your math and the experience to trust your instincts.

Don't compete on price

If you're ever in a situation where you can overhear a couple of design studio heads having a casual conversation, try to listen in. The way each maneuvers to get the other to reveal their pricing structure is as fascinating as watching two fat old bulldogs in the dog park circling each other for a genital sniff, but with much less probability of success. And a lot more slobber.

Here's the weird thing: opacity always favors the buyer. Those studio heads will then compete against each other and drive their own prices down without the buyer ever having to lift a finger. The buyer will end up getting a deal. The "winning" studio will be stuck completing a project on which they're barely making margin.

Sadly, the going rate for design work isn't posted anywhere like the price of apples. And unlike competing grocers, you can't casually walk by their apple stand and see what your competition is charging to adjust your prices accordingly. But with a little savvy and a bit of friendliness, you should be able to get a general idea—at least a ballpark hourly rate—of what others are charging. (Though honestly, we should all just tell each other.) Decide who you are competing against, and you will at least have a good sandbox to start playing in.

There is always someone cheaper. Negotiate price, but don't compete on price. Compete on quality, value, and fit.

Salaries are the same deal. You may notice that companies of relatively the same size offer salaries all over the map for the same level of job. You'll definitely notice that they offer lower salaries if you're not a white male. Say it with me, folks: this industry is sexist and racist as fuck.

Again, we should all just tell each other what we're making. The last thing employers want is for workers to start comparing salaries and coming to the realization that they're being compensated differently for doing the same work. They especially don't want you to notice those differences falling along racial and gender lines. Which is exactly why you should do it! There are sites for salary comparison (Glassdoor being the current biggest example), and it's not too hard to find anonymous spreadsheets of designers sharing what they make.

Hopefully, salary-sharing becomes normalized. If you're uncomfortable with people knowing what you make, flip it around and think of it this way: we should all know what employers are paying us. Get comfortable talking about money, because employers are using your discomfort to exploit you.

Charge for value, not time

If you're charging by the project, there are formulas that claim to help you figure out what to charge by looking at how much you need to make. They'll have you add up your expenses, rent, supplies, utilities, etc., and then add a markup to that based on how much profit you want. But there's a problem with most of those formulas: they do not calculate a billing rate—they calculate a cover-your-ass rate. You should, of course, know the minimum rate you need to make to keep the lights on. But you should be charging clients based on what that work is worth to them, not the time it takes you to complete it.

Let's say you're working on bids for two logos. One is a logo for a family-run sandwich shop around the corner from your apartment. The other is for a VC-backed startup that's been valuated in multiple billions of dollars. And let's also say you expect both of them to take you the same amount of time. (Bear with me as I oversimplify the hell out of this.)

Are you going to charge the same for both? No, you are not. Because the first logo is designed for a family-run business in your neighborhood that has a water bowl outside for the local pups, while the second is for a company that will turn around and use that logo as part of a service that generates a fuckton of money selling your neighbors' data to third-party brokers. (This is every company, by the way.)

Only you know the value of your time. (Hint: it is greater than zero dollars.) But the value of your work to a particular client depends on what the client has to gain from that work. And the client is not buying time from you. They are buying work. The value of that work is directly linked to how much money will be generated from the work. That's what you need to charge them for.

Do your research

This is different from the research you'll be doing once you sign the project. You get paid for that. This is research in support of putting together a smart bid or salary demand.

Find out as much about the company you'll be working for as possible. Do they have a history of launching products on time? Is there a possibility that they'll be acquired or shut down during the project? Do they have a history of exploiting their customers and/or workers? Do they have a history of underpaying employees of color? (Do they even have any?) Are you dealing with the actual decision-makers? If not, can you access the actual decision-makers? And the super obvious but oft forgotten: Do they have money to pay you?

Find out about their internal teams and their involvement in the project. If they're responsible for major pieces (content strategy and the backend buildout are two excellent examples), have conversations with the people responsible for that. Make sure you'll all be speaking the same project language. (This also goes a long way toward building allies in the upcoming project, by the way.)

Do some research into the market landscape as well. Maybe there's a good reason why no one else has been able to sell mismatched socks online. (I wrote that sentence ten years ago. I'm now pleasantly surprised that I've seen ads for multiple companies selling mismatched socks online, which, when you see everything else we've done online, feels pretty sweet and benign. I got a couple of pairs.)

Not only will this research help you prepare a solid price, but the knowledge you gain from it will also make you a better salesperson, to help you to close the deal with confidence. (Or tell you enough to stay away from the project!)

Figure out what the client really needs early

Most clients will approach you with a wish list of desires. (This goes for internal teams as well.) If they don't, work with them on coming up with one. Then figure out which needs are crucial to the project and which would be nice to have, which ones actually fit into their overall goals, and the costs and benefits of each. (It'll be easier to talk them out of the non-crucial stuff once there's a price tag on everything.) Not only will this help you come up with a proper scope, but it'll give you a handy list of priorities should you need to adjust the budget. It also keeps additional things from popping up later in the project, you know, after it's been scoped.

"But it'll only take me an hour!"

Good for you! That should be a highly profitable job.

Don't punish yourself for being good or efficient. I've seen designers pull their hair out trying to solve something, and I've seen designers nail something on the first try. Quite often these were the same designers on consecutive days. In either situation, what the client pays for is that solution, not the time it took you to get there. If it takes you less time than what was estimated and agreed to, and that estimate was created honestly based on previous similar work: great.

(I'll be honest with you, because I love you and you're not going to share this with clients: I have actually sat on work for a few extra days and waited to deliver it closer to the estimated date just to avoid this conversation. But I'm a bad person. I'm certainly not recommending you do this.)

The flip side of that, of course, is that you can't charge the client for the time it takes you to learn a particular tool (unless you're working in-house). If a client hired you to build something on a platform you don't know yet but feel confident you can learn, the time it takes you to pick up that expertise is on you. Once you consider yourself capable, turn the meter on.

Large jobs versus small jobs

When I was in college I took a holiday job loading UPS trucks at night. I trained with a guy named Frank. First piece of advice Frank gave me was, "Watch out for the small boxes, kid."

"Why?"

"You'll find out."

Sure enough, I absentmindedly grabbed the first small box that shot out of the chute and it slammed my hand into the floor. Frank laughed and laughed. He laughed even harder when he realized I needed stitches. I really liked Frank. (The box was filled with ball bearings, by the way.)

The moral of this story is that small jobs contain the same kinds of problems as big jobs, minus the budget and the time to deal with them adequately. You'll try to squeeze them into as little time as possible, and they don't want to be squeezed.

Remember, a small job for you is not necessarily a small job for the client. And while you may be thinking that this is a little something you can fit in between big projects, the client is still expecting the same amount of detailed attention you're giving your bigger projects. And they'd be right to expect that.

GETTING YOUR PRICE

As a kid I was lucky enough to be raised in the fine city of Philadelphia and unlucky enough to witness baseball games at Veterans Stadium (pronounced VEH-tchrn.) Veterans Stadium was one of those terrible multipurpose complexes built in the '70s that tarnished the landscape in cities across America. Anyhow, as a kid I didn't have much money, so my friends and I would get seats in the 700 level, the highest level of the stadium. (Between pitches we'd throw batteries at the airplanes flying below us.)

This is an incredibly long-winded way to tell you that, as soon as a potential client calls, we try to give them what we call a "700-level ballpark figure." Meaning, we can't see too much of the detail on the field yet, but it sounds like it's, say, an \$80–\$120K project. As we talk through the project needs together, we can gradually make that figure more precise, so that by the time they get a proposal in front of them, it's a figure they'll be prepared for. If you're applying for a job, get a salary range before taking on the strenuous work of interviewing. You don't want to be offered a job only to find out the salary won't pay your bills.

If the person paying you is going to have sticker shock, get it over with as soon as possible, before you've taken the time to put together a detailed proposal or revamp your portfolio.

Nothing in your proposal should be a surprise

Some people treat proposals like they are passing a number in a folded piece of paper across a table to be peeked at and then accepted or rejected. Just as a design project requires client input and feedback along the way, so does a proposal. Start with a high-level draft and iterate quickly, with the client pointing out potential roadblocks and offering insight into anything that may trip up their team.

The frank and open communication you begin here will serve you well throughout the project. You need to expose all implicit assumptions because the road to feeling ripped off is paved with assumptions about what you were getting. You and the client may agree on a number but have radically different ideas of what that number includes.

It's perfectly fine to summarize what is included and what isn't, for reference. Another way to handle this is to list everything that needs to be done on the project and who is responsible for it, whether it is you, the client, or a third (even as yet unnamed) party. You compile this list by means of a video call or in-person conversation in which you go through the list of everything that you anticipate will need to be done and ask the client who they would like to be responsible for that work. Everything you are responsible for goes into your estimate. Everything you aren't responsible for goes on a list of scope exclusions.

Here is a short list of terrible surprises to avoid, just to give you an idea:

- Copywriting
- Backend development
- Localization
- Video production and motion graphics
- Custom illustration
- Content migration
- Print collateral
- CMS customization
- SEO
- Maintenance

Your personal charm and enthusiasm for your work should convert your main client contact into a strong ally by the time the proposal is finalized, so that you already have the blessing of a person on the inside when it's time to present the proposal.

Present it, don't send it!

Putting together a proposal can be such an onerous and

time-consuming job that getting it all wrapped up can feel like a victory. And it is—of sorts. But you haven't won anything yet. Getting the proposal ready just means that you're now ready to sell it. It's really enticing to write a nice little email, attach the proposal, and ride off to sit on the couch or walk the dog. It's also a great way to lose a job.

Don't ever leave a proposal on its own. A proposal is a prop for a sales presentation. If at all possible, get yourself in front of the client's decision-makers and sell. Everything you learned while preparing the proposal should be at your disposal. Read the room. Go over as much or as little detail as the room requires. Make sure you go over key benefits of the proposal. You don't sell features, you build features; you sell benefits. You are convincing them to hire you, not to accept your proposal. The proposal is merely one data point toward that decision. Your presentation skills will decide how big a data point it is.

Three things can happen now: you can not get the job, you can get the job right away, or you can be asked to negotiate. It's a little bit like the three bears. And guess whose porridge is just right? Getting the job right away? No.

Negotiation

Having to negotiate your price isn't a sign that something went wrong—it's a sign that something went right.

You will know you charged too little if the client agrees right away. There's nothing quite as disconcerting as seeing a client's eyes light up and their mouth turn into a Cheshire cat-like grin when they see your estimate and their pen races to sign the dotted line before you realize your mistake.

Ideally, your price should require a bit of negotiation. You don't want to charge so much that they don't feel like they can negotiate, but you also don't want to leave money on the table. You want a client to feel like your time and expertise are valuable, that you're right for the job, and that it may take a little work on everybody's part to make this happen. Once a client has decided to put effort into working together, they're invested in the relationship and it's in their best interest to get it to work.

If the client wants the price lowered, go over all the items in the proposal and find out what can be cut. Never lower the price without taking something away. And never take something away without explaining the lost benefit. If the lost benefit wasn't that great, then maybe it's a fine thing to cut anyway. The amount isn't arbitrary; every item has a set cost. So if they want to pay less, they have to be willing to get less.

When I talk about things to cut, I am talking about the amount of stuff the client will have delivered to them, i.e., site features. You can't cut parts of your process that you need to do the job. The answer to "Can we take out this research?" is always no.

Story time: I was dealing with a potential client who kept insisting we cut the research. I gave the usual spiel about how research informs everything. It wasn't getting through. I finally hit my not-caring-if-we-get-this-job place and told him I'd work up a proposal without the research. The next day I showed up with a proposal that included no research. It was twice as expensive and took twice as long. "That's what it'll take to get the same results without the research, Bob." He signed the original proposal.

Remember, the client wouldn't be negotiating with you if they didn't already want to work with you. So negotiate from a place of confidence. That means being okay with not getting the job. Confidence doesn't come from knowing you're right—it comes from being okay with failing.

"Why is your bid higher than the others?"

I love it when a client asks this question. You will learn to love it as well. First off, congratulations on having the confidence to bid high but not too high! Second, your high bid did not scare off the client. Furthermore, you're being presented with an excellent opportunity to continue to sell yourself. The client is asking you to close the deal.

You need to sound confident as hell that not only did you bid correctly, but also you are the best possible fit for this job. And by going into detail about how you got to that price, you can show them that you have a thorough understanding of what it will take to get that job done. Which, by the way, you better have.

Also, this is an opportunity to build yourself up, not tear others down. Don't bad-mouth other bidders; you're not helping yourself by doing that. Quite the opposite. You don't want to get a job because you convinced someone the other bidders sucked. You want to get the job because you're the right person for it.

Why clients think they can lowball you

First off, let's define lowballing as the actions of a client who is not necessarily constrained by budget but wants to convince you that you should undercharge for your work. This is different from a client who values your work but is stuck with a fixed budget.

It's a client's job to get the most work from you for the least amount of money. And they're comfortable doing it because you probably presented the price in an apologetic tone, perhaps even a little embarrassed to be discussing money at all, which actually makes them worried they're not hiring the right person. But again, this is their job, even if they very much want to work with you. It would be financially irresponsible of them not to at least attempt to hire you for less money. It would be financially irresponsible of you to fall for it. Your job is to do the least amount of work for the most amount of money.

Get comfortable saying, "No." Go stand in front of a mirror right now. C'mon. Let's do it together. "No. This is my price. This is what it costs to work with me." If you need some inspiration, think about the times we asked for help with student loan forgiveness and the government said no.

How can you prevent them from doing it? Well, present with confidence and conviction and back your numbers up with good solid research, of course. But more important than trying to prevent them from doing it is how to address it when it happens. Stand your ground. You gave your price. And like I said, it's their job to try to talk you down. They've got to try, right?

You will lose work. But you will also feel better about the work you get. And most importantly, you won't get stuck "winning" jobs with no profit margins. You may even get a reputation as a hard-ass. I can tell you from personal experience that it's not the end of the world. You have student loans to pay.

Negotiating your salary

As a special treat to all the in-house folk who read this whole chapter looking for their deal, I wasn't lying when I said that it all applied to you. It does. But what you're really looking for is what to do with that job offer you just got, right? I got you. I'm happy to give you the insider dirt, because I've been writing offer letters for twenty years. Here it is:

The first offer is bullshit.

Seriously, look, you've just emerged the winner of a pain-in-the-ass battle royale interview process. For reasons both earned and unearned, you emerged at the top of that pile. You're exhausted and you're ready for this whole process to end, for which I cannot blame you. And that's exactly what your new employer is counting on.

Remember that everyone has a role to play here. Their role is to get you on board for as little as possible, and their first offer will reflect that. Your role is to get paid as much as possible, and if you accept that first offer, you're not playing your role. Personally, I've always been surprised—and a little worried— when someone took my first offer. If you're not going to look out for yourself, how do I know you'll look out for the shop during a project conversation?

So, thank them for the first offer, and make a counter-offer. Don't threaten them. Don't mock the initial offer. And honestly, don't go into detail about why you need a higher amount. It's not about need, it's about value. Try something like: "I appreciate this offer, but here's what I'm worth." And remember, while some places may have set tiers on how much you can make, there are other things you can negotiate with: more money, more time off, shorter hours, short salary-review cycles, more options (they'll give you those because they'll most likely be bogus), a bigger benefits package, a clear path to advancement, and more money (yes, I mentioned money twice because most of those other things are bullshit; you want money). Don't negotiate for a bigger title because you'll just be picking up more responsibilities for the same price.

And listen, I hate to keep tapping the Norma Rae sign (I don't, actually), but if you're uncomfortable with salary negotiations, you can get a union job, where the salaries have been negotiated with workers like you at the table, and not set by management, whose job is to pay you the least amount possible.

GET YOUR MONEY

Look, money is hard. It took me years to get comfortable with the idea of being a financial grown-up. And I doubt many of you got into this business for the money. I'm guessing you're here because you love design. But to practice your craft, you need to keep the lights on, and you need your financial house in order. The more attention you pay to this stuff at the right time, the less of your overall day you'll spend worrying and fretting about it. Don't worry about money—deal with money.

The secret to getting the price you want for your work is having done the homework to know you're asking for the right thing, the confidence to ask for it, and the willingness to walk away when you can't get it. Now that I think of it, I just accidentally gave you the secret to life, which I was saving for another book.

5.

WORKING WITH CONTRACTS

I'LL ASSUME EVERYONE READING THIS BOOK GOT THE SEX TALK AT SOME POINT DURING THEIR PASSAGE THROUGH THE STICKY DOORS TO ADOLESCENCE.

The talk where a parent, or auntie, or—good grief—a pastor pulls you aside and attempts to talk to you about sex. And being safe. And respecting the other person. And consent. And protection. And you're pretty sure they're more freaked out by giving the talk than you are by getting it.

Some of us, including me, have now had to give this talk to our own teenagers. (Any doubt I had about her being my kid vanished when she said, "Yeah, I already know all of this from 4chan.")

Talking to teens about sex is a lot like talking to designers about contracts. "We're being careful. We're in love. We trust each other. They have an agile process. He promised there wouldn't be any backend development."

A contract is like a prophylactic. It won't keep you from getting fucked, but it may keep you free from additional liabilities down the road.

See how I lured you into a chapter about contracts with tawdry sex talk? At the end of the chapter I'll tell you a great dirty joke, too.

WORKING WITH CONTRACTS

In the last few chapters we learned how to convince people to work with us, and how to tell if we were selling our labor to the right people. Everything was great. We sold them on what a great partner we'd be; we were ready to start working; everyone was happy. Shouldn't we be designing by now? What went wrong?

Nothing. And we're going to do our best to keep it that way, by specifying the responsibilities of each party to the greatest extent possible in a written document signed by both parties.

And if you're thinking you can skip this chapter because you work in-house, let me reassure you that there's nothing more dangerous than breezing through an employment contract, which we'll be covering, along with union contracts. Plus, as an in-house designer, you might end up on the other side of these discussions if your employer is hiring contractors or consultants, so knowing this stuff might help you have another designer's back.

WHY A CONTRACT?

A contract establishes the nature of the relationship between all parties and makes the important stuff clear to everyone involved:

- What are the ground rules for working together?
- How much money is exchanged and at what point?
- What are the expectations, who sets them, and when are they due?
- What happens if there is a delay?
- What happens when extra labor is needed?
- Who has what rights? What rights does the client/employer get and when? Which do you retain?

And very importantly, a contract states what should happen if everything goes to hell. Almost no one goes into a relationship with the intention of screwing the other party, but things happen. Markets collapse. Funding falls through. Leadership changes. Employees get flushed like dirty water to meet stockholder promises. The CEO decides they're a libertarian and starts handing out copies of Ayn Rand books.

Often, the problem is that one party assumed the other was responsible for something when this wasn't the case—something like a trademark search or content migration. Then there are deadlines, which get missed on both sides. Do you want to be liable when your client or boss tells you missing the holiday shopping season puts their year-end bonus in jeopardy, meaning they might have to forfeit the down payment on their Cybertruck? No, you don't. Because literally nothing could be worse than that.

You can prevent a lot of weirdness—expensive, painful weirdness—down the road if you clarify all this up front. Getting entangled in ambiguity or dealing with a legal dispute keeps you from getting paid for your work, creates a huge amount of frustration, threatens your reputation (your most important asset), and keeps you from spending your day designing stuff.

In short, a contract is what you need to start building trust between all parties. No contract, no basis for trust.

GET A LAWYER (A LOVE STORY)

Many, many years ago, when our company was still in its infancy, we were working with a particular client on a website. And, yes, we had a contract in place, one we'd cobbled together from a combination of online resources and other design companies who'd been kind enough to share their contracts with us. The project was going well enough. Until it wasn't.

The client, for whatever reason, had decided he was dissatisfied with our design solutions. It happens. We work through it. Except this particular client decided to bring in another designer, someone he'd previously worked with on print collateral, to design the site. Since the project was for a full build-out, he expected us to take this other designer's work and build the site. (Small aside here: if anyone approaches you to work on a project they've already hired another designer for, walk away. Professional courtesy. More on that in Chapter 8.)

Our contract specifically stated that the client couldn't bring in another designer to do the work he'd hired us for. So we pulled him aside and talked to him about it. (Contract conversations should happen in person, if at all possible, and never in front of the whole team. Gives everyone a chance to save face.)

Unfortunately the client did not see it the way we did, and threatened to sue us for refusing to finish the project. Not having been in business for very long, and not having huge reserves of cash, we panicked. We'd never had a client threaten to sue us before! Several conversations with the client only seemed to make things worse.

We knew we needed a lawyer, and fortunately we remembered the name of the friendly corporate counsel from a previous employer. She recommended we call her partner's son, a smart young lawyer at a local firm. His name was Gabe.

Gabe listened to our story, looked over our contract, smirking while he read it, and said the most magical phrase I'd ever heard: "I can make this go away." And after a calm but stern phone call, he did.

Sadly, we had to fire the client. (Yes, sadly. I take no pride in ever having to do that.) But we were able to walk away relatively unscathed, with a lesson learned.

Most importantly, we added a valuable member to our team: Gabe Levine is still with us to this day. After fixing our standard contract, he now reviews incoming contracts, gives us client relationship advice, and has even come on stage with me when I talk about contracts.

Why am I waxing so rhapsodic about my lawyer? He

makes sure my contracts are strong and helps me negotiate with confidence. Having him as an advisor makes me confident enough in what I'm doing that I ask for what I'm worth and don't negotiate my rights away. But most of all, I love my lawyer because he makes me money.

By and large the first thing I hear when someone tells me they're having a problem getting paid and I tell them to hire a lawyer is, "They're too expensive." Yes, they do cost money. But so do you! The better our contracts, the more secure our client relationships. His job isn't to sue clients—it's to make sure we never land in a place where we have to. And ultimately he is trained and knowledgeable in things that I am not. Not having to do those things, and letting someone do them well, means I can spend my time doing what I'm good at: designing. I happily write two checks every month. His is the first; my therapist's is the second.

The next comment I hear most about lawyers is, "Are things at the point where I need a lawyer already?" Things were at that point when you decided to stop being an amateur and turned pro. Look: you don't need to carry your lawyer around with you or talk to them every day. Work with them on a solid master contract to cover your most frequent needs, especially if you're a contractor or freelancer, and then get them on the phone if the client has questions about it.

HOW TO INTRODUCE YOUR CLIENT AND YOUR LAWYER

For the most part, your lawyer is an invisible advisor. They should review all of your contracts, but they mostly stay in the background. This isn't to say that your lawyer needs to be a secret. In fact, you should make no bones about telling your client or employer that you're having your lawyer review something. They are doubtless doing the same. Sometimes, as in the example above, just letting a client know that you have a lawyer as well helps to put them, let's say, at ease.

There's nothing wrong with letting your lawyer and

the client's lawyer settle small contract disputes directly. It usually saves a lot of time. Heck, they probably went to school together. But never, ever, and I mean never talk to a client's lawyer directly without yours present. Only lawyers talk to other lawyers. Sometimes negotiating a very large project requires a friendly conference call with everyone and their lawyers. You can accomplish in one hour what might take days exchanging tracked changes.

We've been on conference calls with clients and found out their lawyers were, unexpectedly to us, on the call too. So we calmly stated, "Unfortunately, we're going to have to reschedule this call when our lawyer is available as well."

This can go one of two ways:

1. The client is upset you have a lawyer. This is a huge red flag. This is a business relationship, and good business people keep a lawyer handy to make sure they don't do anything stupid.
2. Or, everyone is fine with it and you reschedule when your lawyer can join in.

Your lawyer is there to help you make sure you're in the best possible position to help your clients without putting yourself in a bad spot. Your lawyer is a defensive asset and not to be weaponized unless absolutely necessary. If you have a lawyer who is turning everything into a fight, then it's time for a conversation, or a new lawyer.

And by the way, HR people are just lawyers who've learned how to smile.

TALKING TO CLIENTS ABOUT CONTRACTS

If you're working with a large organization you may spend some time haggling about the contents of a contract, and they may even attempt to substitute their contract for yours. In either situation, you'll want to confer with your lawyer. However, if you're working with a smaller organization, or even with an individual, they may balk at the idea of a contract at all. And just as it's your job to sell design, it may also become your job to sell them on the need to define the business relationship properly.

If a client flat-out refuses to sign a contract, that's a red flag. But don't give up right away: it may be that they're just uncomfortable with the idea and need some persuading. You'll need to learn to make your case for why the contract is important to both of you, and be ready to walk away from the job should you be unable to convince them. A larger company may also refuse to sign your contract, and ask you to sign theirs instead. (They may refer to theirs as a "Master Services Agreement," the same way Swanson calls hamburger "Salisbury steak.") In which case you should run it by your lawyer before signing.

CONTRACTS PROTECT BOTH PARTIES

Every relationship you have has an implicit contract. For example, if your buddy calls you and tells you that they just got dumped, it's implied that you will comfort them and talk shit about their ex. If your kid is sick, it's implied that you'll make chicken soup. If you're sick, or we're in a pandemic, you wear a mask to protect others. If you accidentally open a neighbor's package, you put it back, along with a note apologizing, and you never, ever mention what you saw in the package, Kyle. These are the implicit contracts that keep society moving along at a nice pleasant clip.

But what happens when these implicit contracts go wrong? Tension builds and, eventually, arguments arise, because one party believes the other party is responsible for behaving a certain way. (This is why I had my lawyer draw up a contract between me and my kid.)

Having an explicit understanding of what is expected from both parties in a relationship—an explicit agreement of what happens should something go wrong—relieves that tension. People go into a business arrangement with the best of intentions and a lot of assumptions. A contract makes those assumptions explicit by documenting the terms of engagement clearly.

If any red flags or concerns come up during the business development process or the hiring process, make sure those are addressed in the contract (unless they are so big and unresolvable that you need to walk away).

For example, let's say you've decided, for whatever reason, to work with a company that has just laid off a thousand employees. Make sure the contract addresses what happens should the team you're working with get laid off. And from the client's side: let's say they're worried about working with an agency that's out of state. Make sure a certain number of visits to their office are baked into the contract, if that's important to them.

That said, both parties don't necessarily need equal protection. It's doubtful that a ten-person studio, or one freelance designer, can bring down a startup of a hundred, much less an established multinational corporation. Make sure your contract protects you adequately.

The larger and more established the organization, the more heinously skewed in their favor their terms are going to be. Always start from a strong position with your desired terms in mind and a good sense of what you are and aren't willing to accept. There will always be other clients and other jobs.

If you've done your work in the business development or interview process, you should have allies in the organization. You should have convinced someone with some amount of authority that you are indispensable to their success. That is influence and leverage. You might hear, "That's just our policy with new employees," but you'll find that there is frequently more wiggle room in policy than you've been led to expect, especially if you have allies on the inside helping you undo the rusted bolts of bureaucracy.

CONTRACT NEGOTIATIONS EXPOSE MISUNDERSTANDINGS

Some of the essential terms—like the estimated price and key deliverables—get fleshed out as you negotiate the statement of work with your client. However, a lot of gray areas may appear during this process.

For example, what does "payment due upon completion" really mean? What is "completion"? Things get ugly at the end when everyone involved doesn't have a clear shared idea of what constitutes "done." Does it mean that the website is live? Or when you deliver Figma files of the final design? When the client is subjectively satisfied with your work? Does the client need to send you written approval?

Settle the terms of the relationship first and save the arguments for the design.

WHAT MAKES A GOOD CONTRACT?

First off, it's good to have a contract separate from a statement of work (SOW). A contract defines the relationship between two or more parties. A statement of work defines a specific project, while a contract can cover a multitude of SOWs. For example, if you've done multiple projects with the same company, you can choose to have the same contract cover all of those projects, but each project needs to have its own SOW.

Approach contract negotiation in a spirit of fairness while being aware of the balance of power. Who is fundamentally in the pole position? It's generally going to be the client. So if you have a good handle on what their anxieties are, you can address them without gutting essential protections.

If you are dealing with a large organization that has enough lawyers sitting around to sue you out of existence in their free time, you need to stick to your guns. Make sure your contract includes all of the following:

- Intellectual property (IP) transfers on full payment. This simple little idea protects you from doing the work and getting shafted. Basically, it means that even though the work may be sitting on your clients' server, until they've paid you for it, it belongs to you. Most clients are eager to get the IP to their work, because until they do, you can make a claim on money they make through that work. Also, you can request it back.

- Termination fee (or kill fee). A termination fee protects you from walking away empty-handed should a client kill the job for reasons that are out of your control. The kill fee should cover the resources you've allotted to this particular client and the fact that you're not out looking for work because you've committed your time to this project. Lots of clients will balk at a kill fee because they think it means that you can walk out on a contract and still make them pay. Not true. A fair contract is binding on both sides. Neither party can walk away without just cause.

  If the client is uncomfortable with the kill fee, then don't allow them to terminate for any reason other than your material breach of the contract terms. As a basic rule of thumb: any contract that either party can just walk away from is not a contract.
- Deliverables acceptance language. This goes a long way toward making clients more comfortable with the provisions in your favor, such as the kill fee. The contract should state that if the client is not happy with your work, they need to let you know and give you a chance to address the problem, but ultimately they can, for all intents and purposes, fire you. That means you keep the work though. Do not get fired and deliver the work. It sucked, remember? Should a client tell you the work is no good and demand the rights to it, they're probably trying to use it without paying you. This work is terrible! And the portions are so small!

Make sure your contract does not include:

- Unreasonable indemnity language. This protects the client from using you as a human shield. For example: say you make some recommendations that accidentally violate a law. Careless of you, yes, and it's something you should fix. But indemnity means that if the client gets sued, they could hold you responsible for their losses. An individual or a small studio simply can't afford the insurance necessary to cover that kind of indemnity language, so avoid it.
- Warranties. While you certainly design for success and test accordingly, you simply cannot guarantee that the design will meet goals. It's like asking a college to guarantee that getting a degree from them will get you a good job. A client once asked me for a guarantee and I told him I'd give him one if we could add a bonus clause to the contract should the goals be met. That was the end of that conversation.

Even with all the right language, though, remember that a contract doesn't keep things from going wrong. Most importantly, a contract can't make the project go right. And while a contract is essential to every project you take on, beware of the false security that might come from statements like, "Don't worry. We have a contract!"

A contract will not turn the wrong client into the right client. A contract is not a parking lot for red flags. If your gut is telling you that a relationship isn't going to work out, trust your gut. It's better to be wrong by passing up a job you should have taken than to be wrong by taking a job you shouldn't have.

ENFORCING THE CONTRACT

Once both parties have agreed to terms, signed the contract, and exchanged copies, put it away. It's there should you need it, but don't use it as a shield during the project.

There will be a million small disputes while you're working together. The vast majority of these should be handled with civil conversations, ideally in person. Handling disputes with conversation will build the client relationship; pulling out the contract at every possible opportunity will undermine it. Save it for those moments when things have actually gotten bad. Most projects don't get that bad. And if they do, it's a sign that you're doing something else wrong.

However, do make sure you understand everything you've agreed to, and act like it. For example, if you have agreed to get approval in writing at every milestone, do that consistently. Don't get lazy just because the project seems to be going well.

If a project gets to the point where you feel like the other party isn't meeting their end of the contract, address it as soon as possible. But approach it with a positive attitude. Assume both parties want to fix it. This should be an account-level conversation—you and the client, one on one. Never whip out the contract in front of either team. You'll have an easier time negotiating your way back to good project health in private. The goal is always to finish the project well.

Should it become clear that you've reached an impasse, then it's time to call your lawyer for advice. Never threaten your client with your lawyer or with legal action until your lawyer tells you to do so.

The more careful you are when you enter into a client relationship, and the more attention you pay to dotting i's and crossing t's, the more likely you are to build a lasting relationship that will lead to new work and referrals for years to come.

In twenty years of business, I have never taken a client to court. I've only ever fired clients twice, and those are none of your business. I've been fired once. In the few times it hasn't worked out, the parting has usually been civil. If a jerk like me can do this, you certainly can. I guarantee you're nicer.

WORKING WITH FRIENDS

The design and tech communities are small. You're going to work with your friends at some point. In fact, there's nothing wrong with getting a chance to pitch on a project because you're friends with somebody on the client team. You still need to sell yourself as with any other client. Nobody should hire you just because you're buddies with Jim in marketing, but being friends with Jim might get you in the room.

The particular land mine I want to warn you about is informal projects with friends—the sort of stuff you feel funny about having a contract for. I'd tell you to avoid these projects, but you'll just ignore me because, let's face it, you'd feel like a giant jerk asking your buddy to sign a contract to design a few pages or make a logo for his pet project.

If you insist on doing these projects, at the very least have a conversation about possible outcomes and detail your decisions in an email to each other. That may be a bit uncomfortable (although it certainly doesn't have to be), but it may ultimately save a friendship.

EMPLOYMENT CONTRACTS

For those times in your career when you decide you want to work in-house, you'll probably be asked to sign some form of employment contract. The bigger the company, the more onerous the contract. And look, I know you're excited to get that job, especially if it was a long job search and the bills are stacking up. So take a moment to celebrate that you've got an offer in front of you. Really let yourself feel it.

Now, take another moment to go over that employment contract with a fine-toothed comb. (Even better, use your lawyer's fine-tooth comb.) Remember that you're about to sell your labor to these people, so get a really good look at what they expect in return.

I'm not saying your new employer is necessarily trying to screw you. Sometimes they don't even know what's in these contracts themselves. They were drawn up by a corporate lawyer ten or twenty years ago and nobody's made a fuss, so they still say things like "employee agrees to wear a hat on Sunday" or "employee is automatically granted membership to the John Birch Society." (You laugh, but a significant number of deeds to American homes still have racial covenants attached to them.)

A good employer should have no issue with you taking the time to review a contract, and if you're about to enter into a long-term relationship with them, this is a fantastic time to find out who they are. If they're pressuring you into signing, they'll be pressuring you into doing other things you're not comfortable with down the road. Only by then you'll have even less power.

If something is important to you, it'll be much easier to bargain for it before the contract is signed than after. They put a contract in front of you because they need something you are uniquely capable of providing, and by their own admission you provide it better than anyone else they talked to. They're not gifting you a job; they're hoping to benefit from your unique labor. You are valuable! So if you want unlimited paid time off, now is the time to fight for it. And once you get it, please take it.

I'm not going into the details of what might and might not be in an employment contract, and what's legal or not legal will vary depending on where you are signing it, or where the company you are signing it for pretends its headquarters is located. But there are two particularly onerous things to look out for in employment contracts:

- Non-compete clauses. This means that if you leave the company, you can't go to work at another place that does the "same thing," with the "same thing" being very vaguely defined in the company's interest. Companies tend to throw these into contracts by default. Beware of signing anything that prevents you from earning a living. If a company insists on a non-compete clause, insist on a severance that covers that same amount of time as the non-compete.

- Arbitration. This means that if shit happens at work, you agree not to take the company to court, but instead to resolve it through a company-chosen third party, who is usually one of CEO Chip's frat brothers from his legacy stint at Dartmouth. Shit happens at work. People get harassed. Racism comes out of people's mouths. Managers proposition their direct reports. When laws are broken, you want to make sure you have full access to the law. I will tell you this again because it bears repeating: HR is not your friend. They don't exist to protect you. They exist to protect the company from you.

And just so we're clear on this, the only reply to "If we do this for you, we'll have to do it for everyone" is "I'm absolutely fine with that!"

GOOD NEWS FOR INTROVERTS

If the prospect of having to deal with all of these contracts overwhelms you, it's because it's meant to. The same way that most folks, even with the best of intentions, will read the first few paragraphs of an app's terms and conditions statement, realize it's ten thousand words long, and quickly scroll to the I AGREE button. Contracts are designed to overwhelm individuals into just signing the damn things, especially when the thing on the other side of that contract is a job! Double especially if you're uncomfortable with the adversarial nature that contract negotiations can sometimes take. Which, again, is by design. Individuals are a lousy match for a corporation.

Until those individuals decide to work together, collectively.

If you want to work in-house, your best bet at being able to do your job to the best of your ability, work ethically, and not have your labor exploited is to sign a union contract.

There are a few ways for this to happen.

The first is a trade union. That's where an entire trade unionizes, and the union goes with you wherever you work. For example, a city might award a construction contract and specify that the developer has to hire union labor. That means every electrician on that job site carries a union card, and the developer deals with the union about things like salaries, breaks, workplace safety requirements, time off, and overtime. Sadly, designers are a few years away from a trade union.

The second, and far more likely, scenario for designers is a workplace-specific labor union, sometimes called a collective bargaining unit. Which means the workers of a specific company decide to form a union at that company. This may or may not include workers with different skill sets. For example, all the workers at Facebook could unionize under—let's say—The Facebook Workers Union, or just the designers at Facebook could unionize under The Facebook Designer's Union. (By the way, Facebook workers are free to use these names if they'd like. My gift.) Workers would stop being in those unions if they left Facebook, whereas a member of a trade union is part of that union no matter where they work, and even if they're out of work.

Either way, all of those employment contracts would now be handled by union representatives, made up of fellow workers, on everyone's behalf. This is called collective bargaining. Wages, time off, bonuses, parental and caregiving leave, etc., would now be negotiated on behalf of all workers at once, and, trust me, you'll have a much easier time getting the things you need when you unite your voices. Also, if you've ever been called into HR because you alerted them to a manager's inappropriate behavior, and they intimidated you into withdrawing the complaint (HR is not your friend), let me reassure you that there is no greater joy than replying to their request with, "I'll need my union rep here with me."

The easiest way to get into a workplace union is to look for a workplace that has already been unionized by some brave souls who worked there before you. And I guarantee there will be more of those places by the time you're reading this sentence than there were when I wrote it. But still not enough. You may end up having to be that brave soul yourself.

Unionizing a workplace can be a terrifying prospect, much like the idea of doing your own dental work. Which is why you don't do your own dental work: there are people who are very good at doing it for you. Same with unionizing. If you're union-curious, it's extremely easy to reach out to a union organizer and have them go through all the steps with you. They're super helpful people and they're workers just like you. They'll look out for you. They'll make sure you're taking all the right steps, and taking them safely. I've put a few resources in the back of this book to guide you.

Most importantly, unionizing does not have to be an adversarial situation. A union doesn't mean you and your boss are now enemies. It means the workers want to be treated fairly and with respect, and are creating guardrails to ensure that happens. Management shouldn't have an issue with that. Making sure that workers are well treated and well compensated is only a threat if you were intending the opposite. Putting on a seat belt doesn't mean you don't trust the driver; it means you're doing everything you can to ensure your safety. But if management sees it that way, well—when someone tells you who they are, believe them.

FREE YOUR MIND, AND...

Look, I'll be honest with you. I'm a worrier. I spent a fair amount of my young life worrying about what I was supposed to be doing or not doing, worrying that choosing to focus on one thing meant abandoning other things I should've been doing, and, worst of all, worrying about what to do if it all went sideways.

After much therapy, I learned the joy of making lists. Big projects went on the list, small projects went on the list, concerns went on the list, phone numbers of who to call if something was wrong went on the list—everything. I emptied out my worries into lists and no longer had to carry them around in my head. I could focus on accomplishing Item A, knowing I wasn't abandoning Item B, because I could see where it was slotted in the list. This gave me peace of mind and room to focus.

Contracts are basically the same thing. You take all the things that could go wrong about a job, all the deadlines, all the responsibilities, all the stuff that keeps you from focusing, and you make a list. A fancy one, to be sure. You deal with it up front, you get everyone to agree that the contract is canon, and now—you can work.

6.

GETTING PAID

THE VAST, AND I MEAN VAST, MAJORITY OF QUESTIONS I GET FROM YOUNG DESIGNERS ARE SAD STORIES ABOUT LATE PAYMENTS, OR SADDER STORIES ABOUT PAYMENTS THAT ARE NEVER COMING.

The saddest example I ever saw was from a designer whose client offered to settle up if the designer would do "just a little more work." I hate to see a designer not being paid for their work, almost as much as I hate seeing a designer not doing the work of ensuring they get paid.

The best way to deal with late payments is to keep them from ever being late. And it takes a combination of good terms, clearly defined agreements, and, most importantly, relationship-building to ensure that checks keep flowing at a healthy pace. It's bad enough to worry about where the rent is coming from, but it's worse to have to worry about it even after earning the dough for it.

Sometimes, no matter how much due diligence you do, you'll still get sideswiped. Teams reorganize, companies get bought, projects get deprioritized, and so on. None of these are things you have any control over, but with the right systems in place, you can not only survive this stuff, but walk away with your check in hand.

Let me tell you a story. A few years ago our team was working on a project with a new division of a pretty large company. We'd done our due diligence. They were solid. The budget was approved in a timely fashion without much haggling, as was the contract. The team we were working with was opinionated, smart, and, to this day, one of my favorite teams to have had the pleasure to work with. A few months into the project, the work was going really well. Honestly, this was one of those projects where both sides really egged each other on to do better work. So much so that they decided to expand the scope of the work, which we were thrilled about. We submitted a change order, which they had verbally agreed to, but not yet signed.

One day we go in for a work session and the place is eerily quiet. We're ushered into an empty office and told to wait for someone to come talk to us. At first I think we're getting shit-canned, but then I realize that wouldn't explain the office being so deserted. We wait for close to thirty minutes. To make a long story short, the company decided to lay off the entire team we'd been working with. It was a shitty thing to do to people who'd been putting their labor into the company. As their senior VP is telling me all of this, my first response is empathy for the people I'd been working with; my second response is to contemplate the vast resources I'd allocated to the project, and the substantial amount of money we are owed. Mister senior VP finishes his story and asks, "Do you have any questions?"

"Not at this time. Thank you."

What followed was a nerve-racking series of phone calls. The first call was to our lawyer, who flat-out told us he'd be lucky to negotiate half our due payment, minus the change order, which was now floating around unprocessed in a corporate no-man's-land. Their lawyer called us and tried to strike a deal, at which point we told him he needed to talk to our lawyer. (Remember, kids: lawyers talk to lawyers. No one talks to cops.) And in an amazing stroke of luck, one of the people the company had laid off agreed to stay on for a few weeks to wrap things up. She reached out to us. Not only did she advocate for us, but she also tracked down the change order and made sure it was processed.

Amazingly, we got paid what the company owed us. And it happened because we'd set up safeguards (our contract), matched their lawyer with our lawyer, and, most importantly, had a strong advocate on the inside. Someone who was willing to go to bat for us because we'd been doing good work together. Without any of those three pieces, I would not have made payroll and quite possibly would have had to lay people off myself.

Before we go any further, let's go over a few safety rules and reality checks. The vast majority of clients pay their bills. Most pay them on time. A small percentage pay late. And very few actually blow town without paying their bills. Everyone gets burned at least once. I've been burned a few times. If I hadn't been, I wouldn't know what to tell you.

The payment process is not so different from the design process in that you start by doing your research.

DO YOUR HOMEWORK

You have more power over how quickly you get paid than you might think. The first step is understanding everything that affects how soon you will get paid. Do your research and pay attention to details. Don't overlook the obvious. And by the way, this is a good list for people looking to take in-house jobs too!

Do they have money to pay you?

In the excitement to take on a project, and under the spell of the client's infectious charm and obvious passion for their project, you may forget to confirm whether they actually have the money to pay you.

Watch out for this especially with startups and serial entrepreneurs—people who specialize in separating others from their money. Which is a fine skill. Just make sure they're doing it to multimillionaire VCs and not to you. "Our seed, first, second, etc., round is about to close" translates to: "We don't have the money."

As excited as you may be about working with someone, don't get yourself into a situation where getting paid depends on someone's funding coming through. That's not your fight. You already have one business to keep afloat.

Where is the money to pay you coming from?

Having money's not enough, though—it also needs to be there when it's time to ante up, and it needs to be earmarked for you. Has the company set aside the budget for this project? Large organizations generally allocate budget before resources, so when GlobalMegaCo comes calling, there's a good chance the budget's already been allocated. But when Joe'sNewNapkinIdeaStartup rings your bell, the money may or may not be there. It's up to you to find out for sure. How? Ask.

There are two budget questions you need to ask during the biz dev process:

1. "What's your budget?"
2. "Has it been approved?"

If you feel weird about asking either of those questions, turn the book back to page one and start again. Keep doing that until you're comfortable talking about money.

If the client balks at answering either of those questions, you have a problem. Usually it'll be the first question that sticks in their craw. Some clients think that if you "trick" them into telling you their budget, you'll tailor your estimate to it. You know what? They're right. If you tell me that you have $200K, and I feel it's appropriate, I'll show you what a $200K design solution looks like. If you have $40K, I'll see if I can come up with a $40K solution. But they'll be different solutions. I'm not going to charge you $200K for a $40K solution just because I know you have it. But the least helpful thing possible is for me to come up with a $200K solution when your budget is $40K. It wastes both our time. So tell me what your budget is, and I won't show you the Audi on the lot when you have Civic money.

The more awkwardly you approach financial conversations with your client, the more awkward inexperienced clients will feel about answering them, and the less inclined more experienced clients will be to trust you.

Is there anything going on at the company that could result in the budget being yanked? Or even the company going away completely? How does their business work? Is it VC-backed, or supported by ad revenue or subscriptions? Is it a foundation sitting on a pile of cash socked away by robber barons a century or two ago? If the client is a VC-backed startup, when was their last round? Are you coming in at the beginning of a round, or toward the end? Does the board need to release the funds?

Some of this stuff is incredibly easy to find out. Heck, they'll just tell you most of it if you know to ask. In fact, asking these sorts of questions often inspires confidence in the client that they're hiring someone who knows how to ask the right questions. Some answers you can get from the news, blogs, or your vast network of connections. You might not get all of the answers you're looking for, but you can get most of them. And you'll certainly get more answers than you would by not asking.

THE INVOICE APPROVAL PROCESS

Once you determine that a company has the money to work with you, you need to find out who cuts the check. This is easier in small organizations, not impossible in others. But don't wait until a payment is late to do this. Do it as part of the project research.

What is the client's standard accounts-payable process? (If there isn't one, it might be a good thing.) Do you need to watch a webinar to understand it?

Remember those movies you watched as a kid where they showed you where the milk came from? They'd start with a rooster crowing as the sun rose, and Farmer Bill walking to the barn carrying two buckets on his way to meet Bessie. They would take you through each step of the process, following the milk from the cow to a pasteurization boiler, after which it was poured into a bottle, capped, packed in a crate, loaded onto a truck, unloaded at the supermarket, picked up by a mom, and finally drunk by a little ginger kid.

Well, the invoice approval process is like that. Except it turned out that Bessie was actually just what he called the incredibly complex equipment developed by the military that pumped a thousand cows simultaneously. And Farmer Bill sat in a windowed office overlooking the facility, his ass in an Aeron, reading crypto newsletters, manning a control panel that told him how much milk was flowing and which cows were underperforming.

Only very rarely, in very small companies, does Client Bill take your invoice to his desk, pull out a checkbook, write you a check, and walk it back out to you. More often than not, your invoice is filtered through an onerous, byzantine, multistep process where at any juncture it can get misdirected, mangled, or simply disappear, never making its way back to your grubby little ginger hands.

It's in your best interest to find out exactly what kind of process your invoice needs to go through when it's submitted, who controls it at which juncture, how many steps are in between you and your bottle of milk, what your options for receiving payment are, and how they might go wrong. The more steps, the more potential bottlenecks.

Just as you design things with actual people in mind

and not for "users," you shouldn't try to get paid by Accounts Payable. You should try to get paid by Ahmed, or Maria, or Stella. As long as you treat Accounts Payable as another faceless process, they will treat you as another faceless invoice. Even the largest, evilest corporation is staffed by moms and dads. Whether bourbon, chocolate, or weed, everyone in Accounts Payable has a thing you can bribe them with to help you get a check in your hand. Find out what it is.

Organization size is the biggest factor

Much like working people leave the biggest tips, small companies tend to understand that other small companies need to get paid promptly. And you're much more likely to be working with, or at least have met, the person directly responsible for paying you. If they don't have a payment process at all, that can go one of two ways: you either convince them to pay you right away, which I wholeheartedly endorse, or your invoice sits forgotten on someone's desk. If they do have a payment process, it's usually just a matter of convincing the person a couple of cubicles over to move you to the front of the line.

Large organizations tend to have less flexible processes with more moving parts and less visibility into the people directly responsible for payment. It's not impossible to find them, mind you. But even if you make friends with Maria in Accounts Payable, which you should, she's not going to be able to change a large organization's accounting procedures by herself. She can, however, help you see into the process. She may also be an awesome resource for little hints and tricks that might speed things along.

Never, ever go over Maria's head to get paid. You have messed with her world and she will hold up your invoices in perpetuity. Plus, it's a jerk move; Maria's a worker just like you. And if you ever hope to work with this client again, you'll forever have an enemy gatekeeper between you and your money.

TERMS AND CONDITIONS APPLY

Getting paid shouldn't be a hassle. You are delivering a valuable service for fair and agreed-to compensation. But sometimes you run into people who enjoy the hassle. Beware of companies that treat paying you as a favor. If a potential client starts making cracks about money, get the hell out of there before it goes further. They're letting you know they don't value what you do. They will resent making payments. You can help decrease the hassle with a few proactive steps.

SETTING TERMS

For the uninitiated, terms refer to the agreement between you and your client about how long it'll take you to get paid. "Terms" is short for "terminal." As in the terminal where you'll be catching a bus back to your childhood bedroom because you agreed to terms that weren't in your best interest and couldn't cover your rent.

Terms are referred to in wonderfully human phrases like net 15, net 30, net 45, and, for fans of comedy, net 60. (We once had a client try to talk us into net 90. It went as well as you'd expect it would go.) The number refers to the number of days that can elapse between the date on your invoice and the date on the subsequent check. And the more institutional the client, the higher the number.

You've probably figured out that you want a lower number, like 15, to pay your bills, while your client wants a higher number, like 60, to keep their money in the bank generating interest longer.

Don't settle for terms you're uncomfortable with. Negotiate. As always, both parties will try to get the best terms for themselves. You propose 15, they counter with 60. So what do you counter with? 30? Who taught you to negotiate? You stick with your 15. Don't blink first. If you counter with 30 right away you'll get stuck at 45.

Remember, they're fighting for an extra 30 days of interest. You're fighting to stay in business. Which they'll need you to do to finish the project. It behooves everyone to get you paid so you're focused on the work and not the finances. Never take on terms that put your business at risk.

Above all, never take on bad terms just to get a job. There are worse things than being poor. Namely, being poor and owing someone your time.

COMMENCEMENT FEES

We are in a special category of industry that gets paid a commencement fee, also called a deposit. It's an industry standard, and don't let anyone talk you out of it. Most clients have absolutely no problem with it. Others get a little weirded out. You know what? They're playing you against your inexperience.

Any industry that has to allocate a substantial amount of resources to create custom work, be they group resources or an individual's time, is taking a risk that needs to be mitigated by a deposit. Design is custom manufacturing. Anybody that's making something specifically for you, and tailored to your needs, is going to charge you a fee up front. What you're creating is not a commodity that can be turned around and sold to someone else. Your tailor, caterer, architect, the dude making uniforms for your little league team, the gearhead putting together your custom bike: all of these people are making things specifically for you. And if you should disappear on them, the pool of people they can turn around and sell those items to is small to nonexistent.

If you go to your tailor or seamstress and get fitted, they're gonna make an outfit that fits you out of material you choose. To minimize their loss and to solidify your commitment, they're going to ask you for a deposit. Now you're both invested in the outcome. They're sinking time and material; you've sunk some money. If your tailor or seamstress is smart, they'll negotiate for another payment from you during a fitting—that way they're covering their time spent on the outfit at different stages. When the outfit is finally finished, you make the last payment and take it home. You also get a receipt, which in our little analogy is the equivalent of the intellectual property passing from the tailor or seamstress to you.

The commencement fee is due upon signing the contract. Don't kick off without it. If you do, you've lost all your leverage and it might as well be net 30. We don't walk into a kickoff meeting without a commencement check. The only exception is when we're working with a large company or organization (read: schools and state) where trying to meet that requirement puts an undue burden on the client. In those situations, we wait until the payment has been processed—but we don't do that very often.

PAYMENT TIMING

In addition to net dates and deposits, you can also negotiate the number of payments. Sometimes lots of small payments are better than a few big ones. And a company may be able to give you better terms if your invoices fall below a certain amount. The larger the project, the more often I like to invoice at smaller amounts. Otherwise you run the risk of going months without getting paid, and each possible late payment carries substantially more risk. You want to avoid a feast-or-famine scenario.

Make sure your payments are tied to clear milestones. "When they're happy with the homepage" is not a clear milestone. Tie it to an event that can be put on a calendar. Clients can argue about whether something is approved or not far longer than they can argue about whether a meeting ever happened. So if your milestone is the final design presentation, you can invoice right after the meeting. Make sure the milestones are under your control. Unless you're responsible for a site going live, site launch is a horrible milestone.

Never tie a payment milestone to a metric. The world is an unsure place. And even when you do the research, the design, the development, and the testing correctly, things that are totally beyond your control can turn the world on its ear. You simply can't carry the brunt of that. As of this writing, there has never been a federal bailout of small independent design studios, and I'm not holding my breath.

Feel free to tack on a late fee for late payment. Make sure it's high enough that you won't be upset if a company takes you up on it, and have it grow incrementally over a period of time.

DEALING WITH LATE PAYMENTS

Ninety percent of dealing with late payments is figuring out how to avoid late payments. So if you picked up the book and skipped right to this section because you've got an urgent problem, just hold on a sec. Yeah, I realize it's urgent. Yeah, I told you to chill a bit. Because I don't want you dealing with a late payment while you're angry.

The first rule of late payments is to remain calm. Under no circumstances should you go screaming angrily into a client's office to get paid. Or start complaining publicly about it on Twitter (X?). Those things may feel like the right thing at the time, but neither of them gets you any closer to your money.

Now that you're calm, you can try to understand why the payment might be late. Most late payments are a matter of disorganization. Very few are a matter of shame. (Ran out of money.) And even fewer are a matter of malice. (They just don't want to pay you.) Until you have evidence to the contrary, assume your missed payment is a matter of disorganization. That's the most likely possibility, and the easiest to fix.

Large organizations might have a strict process, but release so many payments that some fall through the cracks. Rapidly growing companies might have plenty of money but be totally disorganized about who and what needs paying.

TRACKING DOWN PAYMENTS

Just how late is it anyway? And where is it? Let's calmly track it down. Call your client. Remind them you were expecting payment and haven't received it. Ask them to acknowledge whether they've sent it. If they tell you they haven't, remind them it's due and ask them to put it in the mail that day. If they're local, offer to come get it or send a courier. It's worth $30 to get your money. If they tell you they've sent it, ask them for details: date sent, check number, tracking number if applicable.

As for the late payments due to malice, or people just being jerks: these are the moments you save your lawyer for. Make every attempt to collect the amount owed you in good faith. And if that fails, unleash your lawyer. Just realize that he or she will charge you for getting that money, but 80 percent of a buck is still more than 100 percent of nothin'.

WHEN THE MONEY DOESN'T MATCH UP

In a crappy economy, everyone pays late. Prepare for this. Negotiate smaller payments more often, so that each one isn't as big a hit and there's less time between them. Go into saving mode. Stash away as much money as possible. Design services is a land of plenty and scarcity.

There will be times that your incoming cash flow won't be there in time to pay bills. Again, you should avoid this as much as possible by staggering your invoices and having a good mix of large and small invoices. Small jobs that pay faster and big jobs that pay more will diversify your payment schedule so that you have a steady stream of small payments coming in between the large ones.

But lean times will happen. Be ready for it by establishing a line of credit with your bank. Depending on where we are in the economic cycle as you read this, and what you look like in the racist society banks exist in and help enforce, this may be easy or it may be hard. But if you're a business owner and you have a good relationship with your bank, walk in and ask; it's a common service. For good measure, try to make it cover at least two months' worth of your expenses.

CHIN UP, SPORT

Sadly, you will have to deal with people paying you late. We've discussed ways to minimize that possibility, but when things fall through the cracks, you'll need to use your knowledge of the situation to track down your payment. Help yourself get paid. I can't guarantee you'll never get screwed over again, but I'm pretty sure I can help you cut down how often it happens.

And if you remember nothing else, remember this: don't walk away from money that's owed you. Go get it. You worked for it, you put in honest labor, and it deserves fair compensation. Get over your awkwardness with money. You're a professional designer, and professionals get paid.

I won't pretend that these conversations aren't hard. In fact I can guarantee that they are. Being scared is generally a good sign that you're doing the right thing. I promise that the second conversation you have about money will be easier than the first, and the third will be easier than the second, and so forth. Until one day you realize you're no longer scared to do it and you're writing a book about it.

WHEN WORK CAN'T PAY YOU

This is for the in-house folks, but especially for the startup in-house folks. There may come a time when you are called into the conference room, or dial in to an all-hands Zoom call, or get a Slack message, and told that the company has run out of money. They might not be so direct about it. They may use phrases like "liquidity crunch," "challenging finances," etc. It all means the same thing: they've exhausted all their funding options. This shouldn't be a total shock, because it's usually preceded by layoffs, which is when you should've started looking for a new job.

After telling you they've run out of money, they may refer to you as "one big family" and ask you to hang around and work for free as they try to "wrangle some financing." But the truth is they've already tried that, and it didn't work. Those paychecks will never come. Walk away. Grab a plant or a lamp, put it on a nice office chair, and wheel it home. Don't let a capitalist's delusion become your financial ruin.

WORK FOR PEOPLE

I will end this chapter with a story and a warning. I'm writing this in August of 2022. In March I did a couple of workshops for one of the largest corporations in the world. (I guarantee you have their products in your home.) The invoice was due upon receipt. I'm still waiting for payment, despite an email chain that would choke a horse.

In over twenty years of business, we've worked with companies of all sizes. Without fail, large companies will demand net 90 (three months from receiving your invoice) to pay you. They will make you jump through bureaucratic hoops and hit you with all manner of byzantine processes. If there's a snag in the process, they'll wait for you to contact them and ask what the problem is before looking into it. Money, to them, is an abstraction.

We've never gotten screwed for payment or hassled over getting paid on time by smaller companies (I'm excluding start-ups from this definition). Companies run by people who still remember what it's like to pay rent by the skin of their teeth have always paid quickly. There've been lean times when I've had to ask for early payments, and a small company has never needed an explanation. You will always do well if you sell your labor to people who still remember what it's like to be people.

7. UNDERSTANDING YOUR VALUE

A WHILE BACK WE WERE CONTACTED ABOUT A POTENTIAL PROJECT.

A nice one: high-profile client, good budget, realistic timeline. We wanted it. We had an initial phone call with the principals and got even more excited about it. They were smart. They asked us really good questions; we gave really good answers. They told us they were talking to a few different agencies, which never scares us. All-around good first impression.

A few days later the client told us they were asking the candidates to sketch some concepts for the proposed site to help them make their decision. And get this! They even offered to pay for it. Not bad, right?

We said no.

We told them that in order to design the right site we'd have to do our research. We'd have to talk to them about their goals, their content, their brand, how they made their money—all that stuff. And we'd have to talk to their intended audience. We'd have to take a look at the competition. We'd have to think about things like technical constraints, editorial process, content strategy, etc., etc. We needed to understand and define the problem we were being asked to solve. Then, and only then, would we propose a solution.

We told them that our process was why our work succeeded enough times that they'd heard of us. Sure, our work is pretty, but, more importantly, it works well. It meets goals. And we hope people enjoy using the stuff we make as they work toward other goals, be it a customer's goal of completing a task or the site owner's goal of getting people to register.

We told them that if we were just to do some quick sketches without the benefit of discussion and research, the ideas would inevitably be wrong. We'd never be able to guess what was in the clients' heads. And we wouldn't put ourselves in a position where we'd be judged on our mind-reading prowess.

And as I'm telling them all this, I'm kissing the job goodbye. Which sucks, because it was a really good job, and I really wanted to do it. But the only way I could do it was if they understood and respected the value that we bring to the table and how we get there. And if you think it took guts to tell someone this, then I wish I'd had them that day, because I was totally freaked out and scared. But you have to do the right thing because it's the right thing, scared or not.

"That all makes a lot of sense. You're hired."

Thunk.

And the next time I had that same conversation with a client, it was just a little less scary. As we tell potential clients when they ask us what their site will look like: "Oh, we have no damn idea. But we know what the process is for finding out."

Your expertise is what enables you to do good work. You'll develop and tweak your process throughout your entire career. Sometimes because you've learned something new, sometimes because the industry evolves and you need to reflect as much, sometimes because you'll go to work at a company that does things differently than you're used to. All of these bits and pieces you pick up will add up to a weird thing called your value. Your value is what you bring to the table.

HOW GOOD ARE YOU AT WHAT YOU DO?

For a few years now, I've been teaching a workshop called "Presenting with Confidence." One of the exercises in the workshop is asking people how good they are at what they do. It's more nerve-racking than you'd think. People tend to look at me funny, like it's a trick question. It's not. But when I finally get them to answer it, most people tend to be very tentative with the answer. They tell me they're pretty good, or that they think they're pretty good, or that they're okay. Then I tell them the bagel story.

Imagine it's a nice quiet Sunday morning. You get up and decide to surprise your partner by walking to the store for bagels before they wake up. You pick up a half dozen still-warm bagels, along with a tub of cream cheese and lox, because you know what's good. You get home just as your partner is waking up, so you clear off the counter, open the bag of bagels, reach for a knife, and cut a bagel in half.

That's when you realize the tip of your thumb and the rest of your hand are on opposite sides of the knife. You scream. You've ruined Sunday morning, a bagel, and quite possibly your thumb. Your partner, who is smarter than you and also not in shock, quickly wraps your hand in a clean towel, grabs a Ziploc bag, fills it with ice, tosses the tip of your thumb in it, throws you in the car, and drives you to the hospital.

Here, the story changes into a Choose Your Own Adventure scenario. You can either be attended to by a doctor who thinks they're pretty good at what they do, or a doctor who is very confident in their abilities. Choose.

Everyone in the workshop opts for the latter.

In our day-to-day lives we have tons of encounters with trained professionals whose services we require. Some of these people will pilot a plane we're on, some of them will fix our leaky sinks, some of them will sell us fish, some of them will reattach body parts that we've chopped off. We will assign them a value based on how well they solve our problems. We might even return to them, or refer them to other people. But in order to let them anywhere near our problems in the first place, we'll need to trust them. And that usually happens because they're confident. They're calm. They look like they know how good they are at their jobs. And they will tell you as much.

We also tend to trust people who have a strong point of view more than someone who doesn't. (That includes having a point of view about yourself.) So when I stop at my local market on my way home and ask the fishmonger what's good, I'm reassured by how quickly they point to the tilapia or the snapper. When I hire a designer, I'm reassured when they tell me they can solve the problem in front of them. I may ask them what steps they'll be taking to do so, but only because their initial show of confidence was enough that I've decided to extend my trust to them.

So when someone asks you how good you are at what you do, think of how you'd like a trained professional to answer that question if you were the one asking it. Ask yourself what you'd need to hear to feel good about handing your design problem, or your severed thumb, over to them.

Humility is expensive, especially when you have student loans to pay. Doing something successfully is hard. Being confident you can do it successfully over and over? That's why people are willing to pay you.

RIDDING YOURSELF OF IMPOSTER SYNDROME

Believing you're good at something is hard enough. Trying to do that while a parasite in your head keeps telling you the opposite is near impossible. And that's exactly what imposter syndrome is—a parasite. (How did it get there? I have no idea. I'm kidding. This is America. The answer is either sexism, racism, or both. C'mon!) It's an especially onerous parasite because it feeds you nonsense and makes you believe that all the shit that's coming from its nasty parasite mouth is actually stuff that you yourself are thinking. It hijacks your own voice. Screw that. No one gets to use your voice but you. So let's ditch this parasite.

I'll take you through a little exercise from the workshop I mentioned earlier. During the workshop I ask people how they got their jobs. (Students can feel free to replace this with how they got into their program. The temporarily unemployed can refer to the last job they had.) Most people will tell me stories that begin with some version of, "I had a friend at the company" or "I got really lucky," and while both of those things might be true, they are rarely the reason you get a job.

I then ask people to dig a little deeper and talk to me about the interview process for getting that job. (Lord, I've heard some shit. Hiring is broken, but I'm saving that for another book.) Most people describe a scenario with multiple interviews, ranging from one-on-ones to getting interrogated by an entire team. Some of the interviews are over quickly, and some of them last the entire day. These people are grilling you.

I then ask how long the entire interview process lasted, from initial contact to offer letter. The answers range from the same day (very rare) to four months (also fairly rare), with the median being somewhere around two months. Do you know what they're doing for those two months? They are grilling other people.

Then I ask whether the interviewers sounded like they knew what they were talking about. And aside from a few nervous jokers, everyone pretty much agrees that they were interviewed by smart people.

Then I ask if the people they interviewed with are the same people they work with on a day-to-day basis. For the most part, other than the typical HR contingent, the answer is yes. They work with these people every day. Which, of course, leads to my next question of whether these people are good to work with. Amazingly, everyone seems to like their coworkers and to believe they're smart. This is an important piece of data. Hold on to it.

At some point we're all going to have to decide whether we're designers who make decisions based on research, data, and evidence, or whether we're designers who make decisions based on pure imagination and fantasy. I think you've read enough that you can probably tell I swing toward the former, and since you're still reading, I'm inclined to say you might be as well. So let me present the evidence:

- You went through a grueling interview process.
- The company interviewed a pool of very qualified candidates.
- The interviewers were good at their task.
- Your coworkers, some of whom were also interviewers, are good at their jobs.
- You are still there.

At this point, the evidence leads us to two possible conclusions: either you are very good at maintaining a complex level of treachery over an extended period of time, or—and hold on to your butts, because this is out there—you are as good at your job as all those people believe you to be.

I don't know about you, but my evidence-based self wants to point to the latter. It's the only possibility that actually makes sense.

Think of it this way: the people who work alongside you, day in and day out, have a certain amount of work to do. You are there to help carry that load. And while we all have our bad days where someone has to help us with our share, there is a limit to how long other workers will put up with you not carrying yours. If you were not doing your job well, you would've heard it from them by now. Not from management; they don't know what's happening on the shop floor until it's too late. It's the other workers who keep you honest.

So while you are there, working alongside your colleagues, helping them do the work, remember that they already know you can do it. They kicked your tires during all those interviews. They kicked other people's tires, too. They picked you. Not because your friend worked here. Not because they thought you were interesting. Not because they thought you'd be fun to have a beer with. Ultimately, they picked you because they thought you could help them get home on time.

They're earning their paycheck, you're earning your paycheck. Imposter syndrome is not earning a paycheck and it's living rent-free in your head. Begone.

YOU ARE A SKILLED WORKER

Like a plumber, an electrician, a doctor, or a butcher, you are a skilled worker. Both of those words are equally important. Skilled—you do a thing that you had to be trained to do, same as everyone else on that list. (I don't want to meet a natural-born butcher and neither do you.) Worker—you exchange use of that skill, which is labor, both mental and physical (but let's be honest, mostly mental in our case), for money. Take away either of those words and you fall apart.

The reason people hire you is because you can perform a task that requires a skill that you have and they need. The reason people hire you is because your skill set benefits them. And that's fine because you have bills to pay, and selling your labor will help to pay those bills. But here we'll switch to a bus driver metaphor: while people can hire you to drive a bus and tell you where they need to go, you'd be a shitty bus driver if you let them tell you how to drive that bus, especially if they demand you drive the bus in a way that puts people in jeopardy.

However, throughout your design career, that is exactly what you'll run into: being asked to do your job in a way that doesn't fit your understanding of how a designer works. (If you're still unsure about how a designer works, I've added A Designer's Code of Ethics to the back of this book.) You'll be asked to skip steps. You'll be asked to create unethical patterns. (Aside: "dark patterns" is a racist term. Erase it from your vocabulary.) You'll be asked to work too quickly, increasing the odds that mistakes get made. You'll be asked to design shit that is straight-up fraud, even illegal. You'll be asked to "just try something so we can see that it doesn't work." Some of the people asking you to do these things may not know these things are wrong. Some of them will know and just not care.

As my therapist keeps reminding me: you cannot control what people say to you, you can only control how you react. Every skilled professional deals with this. Every bus driver has a passenger telling them to drive faster. Every electrician has a homeowner telling them how to wire something. Every basketball coach has people on the sidelines telling them what play to run. It's part of the job.

And every skilled professional needs to be able to look at the people paying them and say, "I've got this." Which is why I just spent the last few pages filling you with confidence. I think I've been pretty clear that that's half the equation. The other half is your design process.

THE VALUE OF A GOOD PROCESS

People get to where they are in life by following a process, whether they're conscious of it or not. Capitalism, for example, is the process of grinding people down to extract their labor. So is waking up every morning, going to a job you hate, and crying in a bathroom stall before lunch. So is unionizing.

But much like the best umbrella is the umbrella you have on you, the best process is the one you're having success with. Your successful process has also led you to enough good work that people want to keep hiring you to do more work. No one cares what's in your portfolio. They just want to know whether you can do those things again, and for them. That's your value. No one is hiring you because you helped Bob sell pants online—they're hiring you because you have a process for figuring out how to sell things online. Bob is just an example of that. They're grilling you about it to see if it was a fluke or whether it's repeatable.

Don't fall for trendy processes just to say you're using the latest trendy processes. Study them. Incorporate the good bits. Don't take a job without finding out what their process is, and whether you're either familiar with it or confident and curious enough that you believe you could excel at it. If you're working on an internal team, your company may claim that they run an agile shop or some other such thing, but you'll notice that the introduction of every new person to the team brings about slight changes as they introduce a new dynamic or twist. Something they've been doing at their old company for years might be an innovation to your team. The best process is the one that allows the team using it to achieve success, and it can come from a combination of many things.

I started this chapter with a story about a client who attempted to break our process, which would have resulted in the exact opposite of what they were trying to achieve. Their goal was to ensure that they got good work, and they came up with a plan they felt would do that. To them, it was a solid plan. But what they actually needed was someone who could achieve their goal, not execute their plan. They needed to know they were in good hands.

Planning how to achieve a goal needs to be something you, as a designer who is good at what you do, take charge of. That's your job. And if you don't take charge of that, you're going to have a very hard time getting anything done during the project.

In this particular case, I addressed the client's goal by ensuring that they'd get good work because we had a process for getting good work done. I addressed their actual concern, and presented my own plan. They were open-minded enough to listen to it, and I managed to convince them their goals would be better met with my plan than theirs. Just as importantly, I followed up by achieving that goal.

Now, should an employer immediately step off the minute you announce you have a process? I'd be surprised if they did. You need to convince them. When someone hires you, they are hiring your process as well, but you need to sell them on that process as the reason you do good work. And that needs to be done as part of the hiring process, not after.

"If we're going to work together, here's how it's going to work."

"If you want to use that thumb again, here's how we're going to do it."

You're not going to ask for permission to do things your way. You're going to convince people that your way works by showing them how you will use your process to meet their goals. And you'll back this up by showing them how many times it has worked in the past. And every time you manage to do this, you'll have yet one more example to make your case, making the next time you have to do this that much less scary.

Our process works. So will yours. If you stick to it. And fight for it.

HOW COMPANIES ATTEMPT TO BREAK YOUR PROCESS

Companies love talking designers out of the process that got them on the project. Why would someone do this? Anxiety. Anxiety about having spent the money. Anxiety about needing to see what they perceive as "results" as soon as possible. Anxiety at having made the wrong decision. And ultimately, anxiety about their own job security should they have made the wrong decision in hiring you.

When people get anxious they fall back into the terrible habits that make up their comfort places. And, by definition, problem-solving and innovation don't happen in our comfort places.

Throughout a project, you may have to remind a client

or employer multiple times that they agreed to follow your process. You may have to convince them that your process is on target to get them the results they need.

There will be hand-holding. There will be tough love. But above all, you will have to stand your ground and stick to what you know works. You will also need to be flexible enough to alleviate anxiety without putting the project in jeopardy. A good process, like a building sitting on a fault line (like the one I'm writing in right now!), is built to give just enough so that it doesn't break.

As for how companies attempt to break your process, enjoy this short selection of popular favorites.

Start drawing (solving) before fully understanding the problem

This is a no. The anxiety here comes from a basic misunderstanding that they hired you to design, and design is pictures. They're not sure why you're doing this other stuff like research. Ugh, boring. And if you've ever walked into a visual presentation and said, "Today we're finally going to look at design!" you're part of the reason this problem exists, so stop blaming your boss. You need to start every project with an explanation of what designing something looks like and how all the pieces fit together. The minute you start putting "pictures" in front of people, you're going to have to address their reactions to those pictures. Don't put yourself in a position where you have to defend either your own work or their reaction to it before you have the research to know whether it's right or wrong.

Work out of order

You know what the least important page of a website is? The homepage. More likely than not it's a one-off template, it doesn't expose enough of the navigational system, and quite often it's more controlled by the needs of marketing than the site's users. Yet this is the page employers want to see the most. I often describe it as building the roof of the house first and then letting it sit until we go back and build the walls.

I hate starting with the homepage. However, I've found that until clients see it, we can't get their attention on anything else. So, we give them a homepage. And while they debate its merits, we get cranking on the rest of the site. We bend, but we don't break.

Try to do your work for you

The most common instance of this, by far, is: "We've already done a lot of research. So we can just hand it to you and skip that whole phase." And yes, seeing that research will be helpful, but it doesn't take the place of doing your own. The point isn't to do research—it's to understand. It's not a checklist item that we're happy to allow someone else to cross off.

The second most common instance is clients developing competing visuals. That's a deal-breaker. Never put yourself in a situation where you're competing to solve a problem you've been entrusted to solve. Plus, you need the room to try things that may or may not be right without the cavalry being called in.

I love competing to get work, but once I get it, it's mine. The competition is over.

Control or block your access to people

Welcome to the world of internal dysfunction. Bob is your boss and Bob is in a power struggle with Maria, but you need to get information from Maria to do the job, but Bob doesn't want you to talk to Maria because he's afraid he'll look weak, or he doesn't want Maria to know what's going on.

There's no way this ends well. Have a talk with Bob before the project starts. Make sure he knows that you will need direct access to people to do the project. If he hesitates, figure out what the problem is; perhaps Bob just wants to make sure he's in the room when you talk to certain people.

But if Bob isn't willing to give you access to other people in the company, that's usually a sign of a bigger problem. Like perhaps the project isn't really that important to the company, or he doesn't have the support he needs to carry the project to fruition.

Rush

I love having a sense of urgency around a project. But certain things will take the time they will take. Measuring twice and cutting once takes less time than an extra trip to the hardware store. Rushing leads to sloppy mistakes, like launching a site filled with lorem ipsum text (not that anyone's ever done that, of course). What's worth doing is worth doing well, and even while working in an urgent manner, details must be looked after to keep quality at a high level.

And I guarantee you that the client who wants to rush the most is the same one that spent a month or more waffling to bring you on board.

Waffle over decisions

Nothing derails a project faster than indecision, whether it's taking too long to make a choice or revisiting choices that have already been acted on. You either have a team sitting on their hands waiting, or a team working backward to unravel what they've made.

You can protect yourself against this by making sure your client understands the repercussions of waffling. Everything on a project has a cost associated with it, be it time or money or both. People hate wasting time and money, and in many cases don't have the resources or authority to get more of either. So when asking them to make a decision, make sure to also tell them the cost associated with it: "If we get an answer to this by Wednesday, then we can move forward and meet the Friday deadline. But anything beyond a decision by 6 p.m. will push us to Monday, and move all the subsequent deadlines out as well. Which means we'll need budget for an extra four days of work."

Obviously, you want to tailor that statement to the situation. Don't pull out a battle axe when a reassuring tap on the shoulder is enough. Just make sure your employer understands that there's no magic trick that allows for 120 hours of work on the last week of the project.

You can do a lot to show how valuable time is in how you comport yourself throughout a project. Run your meetings and work sessions efficiently. Come in prepared. Don't run over time. Don't hang out chatting. Don't train people to think you've got extra time on your hands. As far as they're concerned, you scoped twelve weeks for this project, there was one right before it, and there's another one right after it.

Waffling endangers your most precious resource.

Ignore project goals in favor of organizational politics

"Because the CEO said so." Every project is subject to one or two peccadilloes from a higher-up that make absolutely no sense to anyone. This is a matter of picking your battles. Some of their requests will have absolutely no effect on the overall experience or success of the project; some of them can be devastating. One CEO we worked with demanded that the site contain a photo of a particular local landmark so users knew where they were located. This was of utmost importance to him, and relatively easy to accommodate. (It's on his bio page, by the way.)

A lot of organizational politics are a consequence of people not feeling heard. This is why you should aim to include as many people as possible in kickoff meetings. Even if they're ultimately not part of the project team, give them an opportunity to be heard. You'll find out useful information that your main contact might have assumed you didn't need to know. You can also learn a lot from noting who your contact attempts to exclude from such meetings.

Ultimately your client may ask you to do something on the project that's detrimental to the project's goal because they've been told to do it, and they're understandably unwilling to defy a higher-up's request. You need to be willing to take that bullet. After all, you don't have to work with these people beyond the project. You can push them a little harder. And you need to be willing to have difficult conversations within the organization that your main client may be unwilling or unable to have.

TALK IT OUT

A job has a thousand moving parts. At some point, one of those parts will break. Something will go wrong. Not necessarily because you did something wrong, or because someone else did something wrong, but because the odds were that it would happen. I've found it best to alert people to this possibility at the very start of the project.

When a project goes off course, make sure to communicate it clearly and quickly. How you handle the communication of something going badly and how quickly you can implement a plan for getting back on track is ultimately more important than the fact that things went off track at all.

At the risk of sounding like a broken record, stay in good communication with your project team and stakeholders at all times. They will accept your process as long as you are showing them results. Make sure to set their expectations correctly as to what is happening when, and keep them informed as those things happen.

Stakeholders will trust your process as long as it's transparent and they can see results—and you're willing to bend a little here and there. Without breaking.

“THE MASTER’S TOOLS WILL NEVER DISMANTLE THE MASTER’S HOUSE.”

—Audre Lord

8.

WORKING WITH OTHERS

I WATCH WAY TOO MUCH TV.

Many years ago, during my company's infancy, I was totally engrossed in a TV show on the Discovery Channel (the monkeys and hammers channel) called Monster House. The premise, in a nutshell: Steve, the foreman, would assemble a group of contractors, each with a different specialty, and they'd have a week to completely redo part of someone's house. During the week, the contractors would bicker, yet somehow, at the end, they always pulled through, and the owners of the house would either be delighted or pissed that there was a fire-shooting fountain in the living room and a moat where their lawn used to be.

But the part of the show I enjoyed most (or, at least, the part that serves our purposes today) was the very beginning. Steve the foreman would gather his whole crew together in a room and go over ideas for what they could do. Everyone worked together on this. The carpenter would suggest a drawbridge in the backyard, the general contractor would weigh in on how to do that while taking out 75 percent of the house's load-bearing walls, and the electrician would estimate how much power to steal from the neighborhood grid to make it work.

The first time I saw this, I turned to my partner Erika and said, "That's how we need to work! We need to involve everyone in the project from the very beginning."

And so that is how we work. At the beginning of every project, all the members of the team get together and throw out ideas. There's a fantastic amount of energy in the room as we simultaneously debate how to organize, design, and build things. As we go, we check on whether someone's idea might take us over budget or blow our deadline, while our lead researcher points out that no one wants a website with a fire-breathing fountain.

That way, when we all head off to our own specialties, we're doing it with agreement on what it is we're going to build.

I've worked at too many places where page schematics were thrown on my desk, fully formed, and comps delivered to developers as a fait accompli, often being seen for the very first time. And I can tell you that it does not work!

If you have the luxury of sharing a room with designers of all specialties, you need to make full use of them. Work together. Early and often. Share in the failures and successes. But I guarantee you: if you find people you work well with, the successes will outweigh the failures.

I don't want to get into specific methodologies here. There are way better people to talk to about that stuff. Suffice it to say that these methodologies all involve (or should involve) working together, listening to one another, having a clear understanding of what your labor is going toward, a shared vision of success, and a visible share of that success to all involved in making it happen.

Depending on where you work, you might have a multitude of other people in the office. Even if you work alone out of your studio apartment, you'll undoubtedly have to interact with other people (other craftspeople!) at some point during the day.

If you're in a small company, many of these roles may be filled by the same person. If you're in a large company, each of these roles may be further broken down into mind-numbing subspecialties. If you're living in a dystopian nightmare, you may never actually meet any of these people in person, and they exist as small little rectangles on your monitor. They're still real, with good days and bad.

They're as responsible for the success of the project as you are. And as deserving of your respect as you are of theirs.

"RULES" FOR WORKING

Throughout your career, you will deal with many different people. I can give you a baseline for how to treat a client or a coworker, but ultimately you're going to have to figure out the best way to communicate with individual clients and individual coworkers. It takes confidence, self-awareness, and discipline to have good working relationships. But if you know what you want out of those relationships, know what the other person's concerns, biases, and anxieties are, and know how each of you fits into the process, you'll do well.

Respect people

You can't work with people you don't respect. Yet I fear that too often we choose not to respect people simply because their ideas, their viewpoints, and the way they approach a problem is different from ours. And that means we're cheating ourselves out of a new (or old!) way to solve a problem. There's always something to be learned from working with another designer, whether they're a grizzled veteran or a fresh face.

Remember those print designers we used to laugh at? Well, the web finally caught up to where everything they knew about layout, color theory, and typography was essential. We weren't doing anything new with design. Our technology just couldn't match what design was capable of yet.

Getting good at doing what you love means having the confidence to recognize what you know, the humility to recognize what you don't, the courage to extend your respect to those who make your faults more visible, the curiosity to listen to their stories, and the grace to thank them for sharing their time and knowledge with you.

Maintain clear roles

To work together without tripping over one another and duplicating one another's efforts, you need to establish clear roles. Decide who owns what. Decide how feedback is going to work. Depending on the circumstances, a communication chain with the client may be in order. And yes, one of you may have to be in charge. (More on that later.)

Hard as it may be, once you decide another designer owns a particular piece of the problem, you need to trust them with it. Communicate often. Give each other honest feedback. If their work is sucking, let them know, and figure out other possible solutions together. But under no means should you "surprise" another designer with your take on something they own. And trust me, as someone who works with other designers, it is so very, very hard to not grab that mouse away sometimes. I'll (think I) know exactly how to fix something, and we'll be talking it through, and my hand starts slowly working its way toward their mouse.

That's a terrible urge. It's disrespectful of the other designer. It's a wasted teachable moment. And above all, it's a sign you don't trust your feedback skills as much as your execution skills.

Remember that the goals of the project come first

Whose idea was it? Who cares, if it's a good idea? If everyone on the team can rally around a common goal, the chances of achieving that goal increase exponentially. If we're spending time debating whether the map transition was Jim's idea or Betty's idea, we're not implementing the map transition. What matters is that we have a good idea to implement. (Caveat: No, Jim, this does not mean you get to claim Betty's idea as your own. In fact, it means the exact opposite. It means that if Betty had a really good idea, you celebrate Betty in a way that helps Betty get rewarded, and makes people listen a little closer when Betty is about to say something. Use your privilege to help others, Jim.)

Whether you are working with other designers on your own team or on a client team, you will ultimately be judged on the success of the overall project, not on individual achievements. So why let ego and pettiness spoil the party? You can learn to feed your ego family-style when you celebrate the completion of a successful project along with everyone else.

When I look back at projects I've worked on, the ones that stand out are the ones where people worked well together, not the ones where individual ideas and achievements stand out. And the halo over a successful project tends to encompass the entire team.

Remember that Michael Jordan won no championships until he learned to distribute the ball and involve his teammates. Once he did, he won six.

Don't drive someone else's route

Every once in a while, we get a potential new client who tells us they'd like to hire us because they're not happy with their current firm. Or they're "disentangling" themselves from their current firm. Or they're not happy with what their current firm is doing and they want us to review their work.

We turn those clients down. You should as well. Never step on another designer's work. Anything you say about that firm's work will be used against them. It's one thing to compete honestly for projects; it's another thing altogether to have a hand in getting another designer fired.

If a client's not happy with the firm they hired, let them

clean it up. You stay clear of that mess.

The same goes for that horrid practice of posting your version of a redesign that has just gone live. By all means, critique it. Write about what works and doesn't work. But keep in mind that another designer was working under constraints and internal politics that you may not be aware of. So write up your thoughts as a user, even as a user who's also a designer. But taking the time to redo their work and posting it publicly is smug and petty. (Yes, it has happened to me. Yes, I'm still annoyed about it.)

Stand united. Or I will yell at you.

Use your privilege

Speaking of yelling... I am a cis white man of a certain age. Probably this isn't a surprise to readers at this point. I've made a good career out of telling people shit they needed to hear, not all of which they enjoyed hearing. And while I've lost a few projects, and even gotten fired for it, I've mostly done pretty well for myself. The word for that is privilege. If you look like me, you get away with shit that other people can't. This happens whether I want it to or not; privilege is a systemic problem and can't be personally opted out of. So, I try to use my privilege to help others. I am also from Philadelphia, and my idea of helping others is to yell at the people hurting them.

For those of you reading this book who look like me and feel the same way, please know that we're the only ones who can get away with telling Chad, who looks like us but doesn't feel the same way, that he needs to stop interrupting people when they're talking. We can get away with calling Chad out for talking credit for Maria's idea. We can call Chad out for deciding the team needs to work late tonight, or for railroading us into hiring his friend Brad, or for planning the offsite at a strip club, or for saying something racist, or sexist, or transphobic. Chad is making us all look bad with that crap, and he's more inclined to listen to us than anyone else (which, duh, part of the problem). We have to call him out because we can. We have unearned and undeserved social capital; we need to use it.

But hey, it takes more than one designer to make stuff happen right? Let's meet some of those other people.

WORKING WITH OTHER DESIGNERS

I took the bus to work this morning. It was rush hour, so I was packed in between other passengers in the very front, right by the driver. Another bus approached us from the opposite direction. The drivers made eye contact. They each lifted their hands to acknowledge each other, as if to say, "Hello, I'm driving a bus, and so are you." The buses motored along. Within ten minutes each of them would pass by another bus and the scene would repeat itself.

Cabbies do the same thing. They'll also allow other cabbies to merge into traffic before them, which they'd never let another driver do, ever. And I imagine that bankers probably have a secret handshake when they run into each other in steam rooms. Just like Masons.

Most of the world's professions share a professional bond. At least enough of one to wave hello as they pass each other and acknowledge that they're members of the same profession.

Designers, however, are another matter. If I am meeting with a client's team for the first time, I can usually tell if they have a designer at the table. (The chunky glasses are a dead giveaway. —Ed.) They're the one staring me down with a smirk on their face. Sizing me up. Probably wondering why I've been brought in to do a job they thought themselves perfectly capable of. This doesn't happen every time, of course. But it happens often.

So often that it needs to stop. Throughout my career I've heard the same complaint from almost every designer I've ever met: "No one values me." And throughout this book I've pointed out myriad reasons why this may be true, and how you've brought a lot of that on yourselves. And how it's within your power to fix it. And I'm telling you this with my heart full of love for you, for our craft, and for our profession. But until you start treating yourselves with respect, you can't expect others to do so. Until designers stop treating one another with the catty competitiveness of contestants in a trashy reality TV show, start supporting one another's efforts, and figure out how to complement one another's skill sets, you cannot expect people to take you seriously.

Until you stop throwing other designers under the bus, no one will ever confuse you with the person driving it.

Better together

Working with other (talented) designers makes you a better designer, and is essential to your professional development, especially early in your career. There's simply no better way to learn your craft than to watch someone else practice it.

But even as your career progresses, being in constant contact with someone else who speaks the same language will make you both better at what you do. Hell, just knowing that there's another designer in the room is sometimes enough to keep you from making lazy choices.

Working with designers who have a different background than you, a different set of biases than you, and a different relationship to society than you is better than working with designers just like you. You'll each see things the other designers in the room might not. You'll be able to augment your decision-making process in a way that designers that are too similar cannot. A room full of Chads will always solve problems for other Chads.

At some point some of you will get into a position where you're interviewing the designers that you're going to be working with day to day. It's human nature to gravitate toward the person who makes you feel the most comfortable. Maybe don't hire that one. Maybe go the other way. Work with designers who challenge you and bring different points of view to the table. (Caveat I shouldn't have to make, but fascism isn't a point of view, or an exercise of free speech. Don't hire Nazis.)

During the (ongoing) pandemic I read too many articles about the Swiss cheese model of hazard mitigation. (Okay, I read two.) I'll sum it up like this: no single thing that we did to mitigate the virus was enough. All of them let a bit of the virus through. Masks were good, but not enough. Social distancing was good, but not enough. Washing your hands a lot was good, but not enough. Vaccines were good, but not enough.

Like Swiss cheese, all of these solutions had holes. But when you put the solutions together, because their holes didn't line up, they worked pretty well. So if you were responsible, you did them all, and it became that much harder for the virus to get through.

Let's map this back to a room full of designers. We all have biases. (We're human!) So we're all going to let some bad design through. But a room full of designers with different biases will be better at not letting bad work through than a room full of cis white boys whose holes are all in the same place.

Peer-to-peer criticism

Having other designers look at your work is different than getting client feedback. It's simultaneously more casual and more intense. After all, these are people who know you more intimately than your clients do. Maybe you work in the same space all day, or maybe you meet up after work and show each other stuff. You're willing to show them your mistakes, your lost causes, your uncertainties; they're willing to tell you when something isn't working, why it's not working, and how you can fix it.

You need to take advantage of these people.

Before this whole "design" thing, I had the pleasure of disappointing my parents, hard-working immigrants that they were, by going to art school. We were taught to express ourselves, bare our souls, develop our own personal vocabularies, and read terrible, terrible French literature. And once a week we would get together and critique each other's work. We'd line up all these highly personal artifacts we'd supposedly poured our souls into, and spend the day giving each other honest, if sometimes brutal, criticism. It didn't matter whether the work was done by a friend or someone you didn't know. The task was twofold: to build both critical thinking skills and thick skin.

But this house of horrors had rules. You could say anything you wanted about the work itself, the effort that went into the work, and the craft of execution, but if you made a personal comment about the artist, the room would stop dead in its tracks. You'd be asked to leave. You simply did not criticize an individual. If you were going to upset someone, which happened multiple times a day, you had to do it because they'd made bad art, not because they were a bad person.

Now mind you, I am not trying to make a connection between art and design. The two couldn't be more different: one is a corporate business tool for manipulating the poor, and the other is design. But there is a connection to be made between those critiques and peer-to-peer criticism. A few months in and something started to happen in those critiques. First off, people left. Having your work regularly criticized by your peers is hard. But those who stayed watched their work get better. And their critical skills improved as well. By the end of the year, we were not only doing better work, but our critical thinking skills were sharp as knives.

Sorry for the art analogy. Let's get back to the office.

I'll concede the point that going into work every day to be criticized by your peers is not very pleasant. And it's certainly not going to work out unless you lay some ground rules. It works if everyone involved respects one another. It works when there's a sense of decorum. You have to talk about the work, not the person doing the work. And you have to recognize that everyone involved in critiquing the work has the best interest of both the project and the designer in mind. (If they don't, you have a bigger problem.)

Start by clearly defining the goals at hand. Discuss whether the goals were met and how well. Even good work can be better. The goal is never to make something good enough. The goal is also never to work hard on something. The goal is always to get it right. (Do not confuse someone having worked hard on something with it being right!)

As the person whose work is being critiqued, you need to realize that the feedback is not about you, it's about the work. And you need to be open to good ideas that come from places other than your own head. There's a balance between defending your work and remaining open to better ideas that take a long time to develop. It takes confidence, intelligence, and an open mind to allow others to help you make your work better. It takes a thick head not to.

No more catty competition

The perception (or reality) of competition poisons the working relationship between designers.

As designers working in client services, we're often brought into situations where there's a designer on the client team. You'd be amazed at how often their existence is hidden from us. I'm not quite sure who the client is attempting to protect in these situations: us, the in-house designer, or (more likely) the client themselves.

We're often brought in after a job has been attempted

internally. This is a tough situation for everyone. The internal designer might feel threatened that someone else is now in charge of a job they couldn't do. (Or rather, they didn't have the support to get it done.) Sometimes the internal designer has lobbied not to have an external team brought in, and lost. Also awkward.

Funny story: we were about a month into a project with a client, redesigning an online service. I'm sitting in the client's office and we're batting around a few ideas that have come up. I mention a particular flow I'm considering. "That sounds like something Sam came up with a while back," someone says.

"Who's Sam?"

"He's a designer that was on this project for a few months. He had a few ideas. Other stuff came up though."

"Can I talk to him?"

"Knock yourself out."

The next day I go looking for this Sam. He digs around in his desk and pulls out a full set of flows and page schematics for the product we're working on. And there were some very good ideas in there.

To make a long story short, we got Sam involved in the project, and a huge percentage of what we ended up designing had its genesis in those wireframes of his. Rather than reinventing a solution from scratch, we used his vast knowledge of the product and ideas that he had already explored and took them further than either of us could have done separately.

And finding out about him was total happenstance.

If you're on the services side, make sure to find the client's internal designers. Their institutional knowledge will be a fantastic resource to someone coming in cold. If you're an internal designer and a design team is brought in, reach out to meet them. Their status as outsiders will make it possible for them to navigate above the company's internal politics. Swallow your pride. You can both do things the other can't. Together you're a stronger team than you are apart.

Sam could have been totally pissed off that someone else had been brought in to work on a project he'd done some really excellent thinking on. We could have been threatened that Sam had not only already done all of this work, but that our thinking was taking us down the same path. Instead, we realized we both understood the problem from different vantage points and decided to tackle the solution together. It helped, too, that the company was open-minded enough to let us do so.

Whether you are thrown together on the same project or accidentally discover yourselves on the same project, you're all better off uniting behind a common goal than you are competing for attention and resources.

My first art director

More often than not, though, the situation above plays itself out differently. Designers are a maddening combination of competitiveness and insecurity. For every story I have about someone like Sam, I have ten stories about walking into a client's office to find them looking at a competing comp from an internal designer. And I'll freely admit to you that, before I knew better, I was the one making the competing comps when my own boss hired from the outside. Why did I do this? Easy. I felt like I was better than the designers that had been brought in. And, more importantly, I was afraid I wasn't.

Like all animals, we crave validation. I didn't care what my boss thought of those other designers as long as he thought better of me. I was very lucky to meet a designer named Matt Lynaugh early in my career. Matt was brought in to do an annual report that my boss didn't think I was ready to do. (For what it's worth, I would have nailed it, dick!) From the moment Matt came in, I was irritated. He could sense it;

I'm not very good at hiding when I'm irritated. So one day Matt invites me to lunch and we start talking design. He starts asking me about the company. And starts running ideas for the annual report by me. Next thing you know, we're working on it together. He's giving me tasks. We're giving each other feedback. He was, for all intents and purposes, my first art director. And he taught me what we could accomplish working toward the same purpose, rather than trying to undermine each other.

So now when I walk into a situation and find another designer in the position that I was in, I reach out to them. Like Matt did. We are all links in a continuous timeline of a shared craft. It is our responsibility to keep that timeline strong. To add to it. To pull the next link aboard.

Act like it's within your power to improve your relationship with other designers, and you usually can.

THOSE OTHER PEOPLE ON THE TEAM

A lot of this book has been about protecting yourself from other people. (The rest of it has been about protecting you from yourself.) But throughout your career, you're going to have to learn to deal with a multitude of other people to get your job done. Whether you work by yourself or as part of a team, knowing how to communicate with and listen to the other people around you will be a key factor to doing your job well.

Working with other people is simple: listen to what they have to say, be clear about your own needs, and find common purpose. The rest is a rounding error.

Project managers

You know what's great about project managers? They manage projects. Just as you're responsible for the quality of a project, your project manager is responsible for getting it done on time and with the maximum amount of profit. This doesn't mean you're not both thinking about those things. It means you each own your part of the project.

This often leads to tension, as your ultimate goal is to do good work, and the project manager's ultimate goal is to do the work on time. And that's pretty much how it should be: with each party pushing for the piece they own, and playing against each other, you can deliver excellent work on time.

The project manager also serves as the voice of the client in the room. They make sure the client's requirements are being met, and do most of the day-to-day checking in with the client. But project managers are not your caretaker. They're not there to manage you, outside of how you fit into whatever project they're managing.

The best way to work with a project manager is to be clear about what each of you is responsible for and to keep them informed. Don't wait until you've missed a deadline to alert your project manager to a problem. As soon as you get an inkling that you're going to blow a deadline, let them know. Help them help you by giving them enough time to triage the problem with the client.

There may be times when you feel like you and the project manager are working in total opposition, but you're not. You're working toward the same goal—you just have different ways of getting there.

You can also make their lives easier (and by extension, yours) by cutting the "creative" routine. Be realistic about your time estimates, mind your deadlines, understand the scope. Be someone who can be counted on.

Researchers

A good researcher will make sure you're not building houses with basements in a flood zone. They run interviews with the client's team and with potential customers. They review the client's analytics. Their job isn't to tell you what to design; after all, no researcher could have told Henry Ford that people wanted cars. But a good researcher could have told him that people enjoyed going faster.

Much as the project manager serves as the voice of the client in the room, the researcher functions as the voice of the user. A good researcher will talk directly to the people the work you're doing is intended for and find out about their habits and proclivities. You should make it a habit to sit in on as many of these interviews as possible. Good research is invaluable in knowing whether you're headed down the right design path.

A good researcher isn't prescriptive; they just gather and analyze the data. It's your job to figure out what to do with it. Frequent check-ins with your researcher throughout the design process can be incredibly helpful for validating your decisions (and steering you away from the wrong ones).

The success of every project depends on how well it resonates with its intended audience. And your researcher is your eye into that audience. Treat them well.

Information designers

An information designer, or information architect, is a type of designer. You may even be one! They figure out the structure of things: navigation, taxonomies, organizing principles, and categories. Basically, they figure out where things live and how you get to them.

Information designers flesh out the structure and underlying principles of the site, i.e., the blueprints (thus the architect moniker). A client should sign off on the underlying structure of the site in a bare-bones wireframe before being shown a presentation layer; that way, we collect agreements at various stages along the way, and things can only unravel as far as the last point of agreement. Like tying knots in a rope.

This doesn't mean the visual designer is sitting on their hands until the information designer gets that sign-off.

For years, I worked in various places where information designers would work away on wireframes, headphones on, uninterruptible, and then drop a hundred or so wireframes on my desk. Depending on the information designers, these wireframes would be in several states of fidelity, from pencil sketches to fully fleshed-out page mockups. They would then disappear into their next project. This is known as "waterfall" development. Because you'd want to put them in a barrel and throw them over a waterfall.

My preferred method of working with information designers is to set up in front of a large whiteboard and diagram what we're working on together. The earlier you start developing solutions in common, the better the project goes. Solve together, execute apart.

Content strategists

We're now far enough into the book that I believe I can give you some bad news: almost no one is coming to the sites you work on because of your excellent design work, and the few that do are there to steal. People show up for the stuff. Design makes the stuff easy to find, and a pleasure to use, but it's not the stuff. The web is made of content. Design is what holds all that content in place.

Don't feel bad; the Golden Gate Bridge is only there to help cars get across, but it's the bridge that people photograph.

Content strategists help clients organize and plan for all the stuff. If you have a content strategist at your disposal, start working together early and often. (That seems like a repeating theme. Weird.) You can't build a container for something if you don't know what size it is.

There are two holy questions of any type of design: Who is it for? And what are you putting in it? Content strategists are a godsend in figuring out the latter question. And that answer needs to inform what you're designing. You can't shove a tiger into a cat carrier.

Developers

If you're a web designer, you should know how to code. However, if, like me, you have the luxury of working with fantastic developers, you may find your own hands-on skills getting rusty, even as you stay on top of all the ridiculously amazing advances in our field over the past few years.

Of all the other people in the office, I work closest with the developers, because until we start making actual code, we're looking at paintings of websites. We work back and forth at a pretty fast pace, riffing on each other's work as we go. I don't hand them things to build; we work on things to build together. The sooner we get into code, the sooner we can start iterating.

For example, just this week I started mocking up the full-width desktop version of a site, and as soon as we'd agreed on the basic framework, my developer, Jim Ray, took that and started working through the responsive stuff. But every fifteen minutes or so, one of us would have to adjust what we were doing because the other one had either found a problem or a better way to do something. We were making quicker and better decisions because design and development were informing each other. Had I attempted to mock up all of those responsive states and then hand them to Jim to code, those mistakes would have been baked in, and we would have spent days trying to chase them down. Not to mention that we would have probably asked the client to approve the final design, and it would have been broken.

We often comp just enough to figure out what it is we're building, which is why we don't include visual comps with our final deliverables. They're a mess, often an unfinished mess, which may have little resemblance to what we ended up building. Don't spend time updating paintings when what the client paid for was a website.

Some people also refer to developers as frontend engineers. I pulled developers out separately because they work so closely with designers, and in many cases the designer is the developer, that I think the relationship is different. Although the more a developer starts moving into programming languages such as JavaScript, Ruby, and Python, the more likely they'll start referring to themselves as an engineer.

Engineers

Engineers build the things you design. They come in many flavors, such as application engineers, web developers, and software engineers. They're often bundled together as back-end engineers.

During my days in the startup salt mines a long, long time ago, I was working on a product redesign for a company I'd just joined. The design team was working through a new signup flow. I was arguing that step three needed to come before step two. (I can't remember if I was right anymore, but let's assume I was.) The rest of the team, who'd all been there longer, were arguing that the steps couldn't be swapped because it wasn't engineered that way. I hate hearing a designer argue against something not because it's right or wrong, but because it means a possible hard conversation. Hate it. So I said, "Let's go talk to engineering."

"You can't!"

"Why not?"

And I realized no one on that team had ever had a discussion with anyone in engineering about a product while they were still designing it. Engineering would build based on a final mock-up, often revising design decisions on the fly because of constraints we didn't know (or hadn't asked!) about, and both fiefdoms coexisted peacefully.

The next day I invited the lead engineer to lunch, and afterward asked if he could swing by my desk so I could show him something. I walked him through the new signup flow, with step three before step two.

"That's not how we do it now," he said.

I explained that I thought the new flow would lead to a higher conversion rate because it moved the credit card fetching to the end, front-loading the rest of the user data, so customers would have more incentive to stay in the flow and be slightly less inclined to bail. He agreed it was a good idea. We presented the idea to the head of product together.

And from that point forward we had regular check-ins with engineering. And it was much less likely that design would be changed during build, because we were catching and avoiding problems together.

Seasoned engineers tend not to go for trendy ideas, and have a lot of experience making pragmatic decisions. They are very good at what they do, just like you. And you'll find that if you explain your rationale for your design decisions—as you need to be able to do with anyone you work with—they will be an incredible resource. But if you're both on separate sides of the gym thinking the other is weird, no one gets to dance.

There's a tendency for designers to think of what they do as "hard" because it's so subjective, and of what engineering does as "easy" because there's a "right" answer. But I can assure you that there's as much, if not more, creativity in how an engineer solves a problem as in how a designer does.

Marketers

I've always been one to believe that all design is marketing. A well-designed chair summons your ass to it, a well-designed bike makes even the most honest among us want to steal it, and a well-designed website makes you want to use it. So why does "marketing" leave such a bad taste in designers' mouths? Well, because most of it is bad.

There are good marketers and bad marketers, just like there are good designers and bad designers. A good marketer works for your audience; a bad marketer works for your advertisers. A good marketer will work with you to put together the best possible experience for your customers.

Their main concern is making sure that the site stays in business so those customers can get their needs met. The easiest way to tell if you're working with a bad marketer is to ask them, "Can you please explain what that means?" after they've said something. If they can't, put a stake through their heart.

Put a bad designer and a bad marketer together and you will get exponential crap. You'll also get a ton of complaining on both sides. And it will all boil down to "this other person isn't making me good at my job." Put a good marketer and a good designer together and they'll achieve things neither of them could by themselves.

So as designers I'd like to implore you to do a few things. First off, stop having a knee-jerk response to marketing. Seek out the good ones and work with them. Second, accept the fact that you also work in marketing. You are creatures of persuasion by nature. Just persuade people to do better things. Start with yourselves.

If you're tired of marketing $200 basketball shoes to poor kids, then figure out what it is you want to persuade people to do, and go do that. Get a good marketing person to help you.

VOLTRON WAS A METAPHOR

At this point in the history of UX design or whatever you wanna call it, there is very little that you can do by yourself. You will have to work with people. Some of those people might have the same skills as you, and that is great. Some of those people will have very different skills than you, and that is even greater. You don't have to like them all, although life is easier if you do, but you should definitely learn how to at least get along professionally. These people will make your work possible. They will see things you don't, and vice versa. They may get on your nerves once in a while, and I guarantee you're getting on theirs just as often. And in spite of it all: you're all workers achieving something you can't achieve by yourselves, and that is beautiful. Love is love.

HR IS COPS!

9.

BETTER WORKPLACES

WORKPLACES ARE CHANGING.

Workplaces have changed before, obviously. Sometimes change is a slow, steady, barely noticeable evolution. Sometimes change feels like a kick. The time we're in right now, as I write this in August of 2022, feels like a kick. And as much as we talk about how the pandemic has changed things (and of course it has), in the case of the workplaces where folks like you and I tend to sell our labor, I think a more honest statement might be that the pandemic has exposed the rot.

And while I am loath to use the word opportunity in close proximity to the word pandemic, change in the workplace tends to come about as a response to crisis, and I would consider the pandemic, as well as our response to it, a fucking crisis. So, possibly, by the smallest sliver of margins, the pandemic is an opportunity to clean out some of that rot.

A few weeks ago, I was talking to a friend who's been working from home during the pandemic (as have a lot of privileged people, myself included). That morning she'd gotten a message from her company's CEO that it was time to reopen the office. They had a plan for slowly integrating back into the office a couple days every week, increasing that number over time, until eventually everyone was present in the office full-time. This is the same method we use to cook lobsters. (By the way, this is the first of two crustacean metaphors in this chapter. You're welcome.) The message from the CEO contained words like mandatory and non-negotiable.

She was understandably anxious about returning to the office. The pandemic isn't over (I myself tested positive just two weeks after that conversation), and the idea of being in closed conference rooms and cubicles shoulder-to-shoulder with other humans wasn't great. The fact that the email didn't contain any information about testing or vaccination status or any safety protocols didn't help. Basically, it was time to go back to the office because the CEO said it was time to go back to the office.

I asked her what she was going to do and she told me she wasn't sure. She then added that she loved her job, she was "fucking good at it" (I can vouch this is true!), and she just wanted to work.

I'm sure you've heard similar stories, or you will soon, as the call to "return to work" will come in waves, followed by the latest variant, followed by another call to "return to work," and so on and so on. These stories may or may not be compounded by having children, partners, and immunocompromised folks at home.

And this is all happening because management wants a "return to normal," back to a time that was very reassuring to them, a time where they knew how to operate, a time where signing a twenty-year lease on a huge space was fiscally sound, a time that worked out pretty fucking well for them because they understood the rules—mostly because they set them.

But like I said, workplaces are changing, and they have changed before. What I left out, however, was the most important part: every time a workplace has changed for the better, it was initiated by the workers.

You are a stakeholder in your workplace. It is your labor that generates the wealth that powers those workplaces. In this chapter we will explore ways to make your workplace better, whether you work in the office or remotely.

IT'S NOT MONDAYS YOU HATE

If you happen to scroll through Twitter (yeah, I'm not calling it X) on a Sunday evening (which I don't recommend because it's full of Nazis and lonely white boys trying to convince you crypto is real), you'll see folks tweeting about how anxious they are about the upcoming work week, a feeling sometimes called the Sunday scaries. You'll also see plenty of memes about hating Mondays, some including Garfield and some not. But the effect can be pretty chilling.

The deal that our Socialist ancestors made with the Captains of Industry was very clear: we sell you our labor Mondays through Fridays, and the weekends belong to us. Both parties have reneged on that deal. Plenty of people reading this use their Saturdays to catch up on work from the previous week and spend their Sundays stressing out about the week to come.

I once asked my therapist what the difference between anxiety and stress was, and she put it clearly: anxiety comes from within and stress comes from without. I'll give you an example. If you're heading on vacation and decide you need to pack ten books about design because you feel guilty about relaxing instead of getting ahead in your craft? That's anxiety. It's coming from within. If your backpack snaps from the weight of all those books? That's stress. The stitching didn't break on its own. It broke from the external weight of the books.

The Sunday scaries? That's not anxiety. That's stress. It's coming from the weight of previous traumatic workweeks.

The greatest trick capitalism ever pulled was convincing you that the trauma it was causing you was a problem inside your own head. It's a problem born from dealing with bullshit in toxic workplaces. And if you're thinking that you've never worked in a toxic workplace, you might be part of what's making it toxic, Chad.

I mention this because the first part of solving a problem is to adequately call it what it is. And with every newspaper article I see lamenting "the problem with today's workers" or "why doesn't anyone want to work anymore," we're being actively gaslit. The problem isn't us.

If we are going to create better workplaces, we need agency. The first part is to acknowledge what is actually broken. It ain't us, it's the workplace. The second part is to understand our role within that workplace. We are what makes it possible. The third part is understanding that, for the people in charge, it's already a great workplace. It works well for them, so they're not in a hurry to change shit. No one is looking out for us but us.

It's not Mondays you hate—it's capitalism.

THE MYTH OF INDIVIDUALS

The modern workplace is generally divided into two classes of people: management, and what management calls individual contributors (or ICs).

You'll notice that the term management contains multitudes. It can be as big or as small as any situation requires. If Manager Bob is taking the team out to lunch, he is expected to foot the bill because he is Management, and he would like to be thanked for footing the bill. If a new unpopular policy is handed down, it will come from Management, only this time the term will include as many people as possible so no one has to carry the burden of blame individually.

You, dear reader, have accepted the role of individual contributor. The word individual isn't there by accident; it's a reminder that management would prefer that you see yourself as having exactly as much agency as one person can have: responsible for the actions of one, doing the work of one. (Except when they give you the work of three.) And, most importantly, yielding the influence of exactly one.

The singular you is very manageable. If one singular individual contributor turns into a problem, or starts asking questions that management doesn't want to answer (such as "Why are there no people of color in management?"), well, it's very easy to replace individuals.

Your power in the workplace is derived from the plural you and not the singular you. Look around. Look at the other workers. See yourselves as the collective you are.

CRABS IN A BUCKET

My grandfather was a man of questionable ideas carried out under the guise of learning experiences. Among these were the ideas that anything worth knowing could be learned at a county fair, a child should know how to siphon gas, and—for our purposes today—the only way to buy fish was at a small town coastal fish market at 5 a.m. as it came off the boats. Mind you, I'm not saying he was wrong about any of these things, just that you don't tend to fully appreciate them as a child. It was during one of these trips to the fish market that I stumbled onto a large industrial-sized bucket of crabs. I watched for a while as the crabs climbed over one another trying to escape the bucket. Just as it looked like one was about to make it to freedom, a crab below would reach out and drag it back into the bucket. "Grandpa, why don't they put a lid on the crab bucket?"

"Don't have to. The crabs pull one another down. In fact, the easier you make it look for them, the more vicious they get about pulling one another down."

"Is this a metaphor?"

Fast forward a few decades and I'm talking to a friend of mine who works in tech. We're about a year into the pandemic. She's telling me how exhausted she is and that she really needs some time off.

"Don't you have unlimited paid time off?"

"Yeah, but if you take any everyone looks at you funny." Crabs in a bucket.

When vacations were doled out as a set amount of time people took them. Two weeks, three weeks, a month for our European friends, it was easy. Everyone scheduled theirs, and when it popped up on your calendar the man took his foot off your neck and off you went on a modest little trip to the coast to look at crabs. There was a lid on the bucket, and you knew exactly when it opened and for how long.

At some point, capitalism visited a fish market, spotted the bucket of crabs, and invented unlimited PTO. Now management doesn't have to track anyone's vacation time because everyone's afraid to take it so they don't get side-eyed by the other workers.

Don't get played like that. If someone hands you unlimited PTO you take it. And if your coworker says they're taking a long weekend, the only acceptable reply is "That sounds amazing. I'm going to take one, too!" Beware of policies that pit workers against one another. Work together and you can all make it out of the bucket.

ON WORKING REMOTELY

At the moment I am writing this sentence, we are two years and four months into a pandemic. By the time you read this sentence it will be longer. (Some of you may be reading it and asking, "Oh, was this written on Old Earth?") At the very beginning of the pandemic, many of us set up shop in our homes. We were the lucky ones. We were able to do our work, and earn our pay, from the relative safety of our apartments. This worked out better for some than others, and trust me that I'm not making light of the moment. But remote work, which once seemed like an outlier that a few designers we knew did, became something every designer did, and will probably end up becoming something that a lot of designers do.

There are positives to this. For one, people aren't driving three hours a day to go sit in a cubicle staring into a laptop that also made the three-hour drive. That's a net gain for the planet.

There are also negatives to this. When you're working from home, you're not just home all the time; you're also at work all the time. And when workers who used to be tied to a specific geographic location started dispersing across time zones, companies seized the opportunity to make the workday fungible as fuck. If a few people on the team are on the east coast, it's only fair to start scheduling meetings at the start of their workday. If a few people on the team are on the west coast, it's only fair to keep scheduling meetings at the end of their workday. The remote work week is fucked. Remember: your boss' job is to extract the most value for the least cost. We keep expecting the parable of the scorpion and the frog to end differently, and it never does.

As a remote worker, it will absolutely fall on your shoulders to keep track of when you are working and when you are not.

Set your hours. Let your manager know those are the hours you're available to work, and then stick to it. Lock your laptop up at the end of the day. Don't check work email. Don't look at work stuff.

Your manager will attempt to get you to make exceptions. That is their job. Don't be upset at someone for doing their job. Do yours instead. Which is to hold the line.

"Did you not see the message I sent you last night?"

"I saw it when I logged back in this morning."

"I expect you to be available whenever I send a message."

"Let me grab my union rep so we can continue this conversation."

C'mon. You knew that was coming.

WHO CAN AND CAN'T HELP

I'll cut to the chase: the short answer to both is you. But you is a funny word—it's both singular and plural. Singular you can't do much. Plural you, which is oddly made up of many singular yous, can help a lot. That said, you will have both allies and enemies in this effort. Let's meet them.

Management

Many a laid-off designer has spent their last hour at work filling a sad cardboard box with "Love where you work" swag. Never love where you work. Work will never love you back. Within our system of capitalism, management's sole concern is to create value for shareholders. Everything else they do is in service of that goal. Hiring you? Keeping you motivated? Telling you that you're all a big family? Building you a big campus where you want to hang out with your friends (as long as they're also your coworkers), providing you with an amazing cafeteria, onsite massage, haircuts, laundry, and, dear lord, company-provided therapists?! It's all to help you create value for shareholders. This is not a cynical take. This is an objective take.

So do your job, and do it well. Enjoy it even. Do you like your management team? Fine. Some of them can be very nice people. But never forget what this relationship is. You are trading labor for money. These very nice people will sack you in a heartbeat if it helps put a good spin on an earnings call. That's their job.

Human resources

The human resources department is not a place of honor. The people in HR will present themselves to you as being "on your side." They are not. Their job is to keep the company out of trouble. If you go to them with a problem, their job is to figure out if this problem will be detrimental to the company; whatever trauma it may be causing you is none of their concern, although they will say that they're looking out for you. They are a wolf in sheep's clothing. If your manager is harassing you, HR's job is to protect the manager. But Mike, I know a very nice person in human resources. Sure, and I met a decent cop once. Individual merit cannot overcome a harmful institution.

Chief happiness officer

I can almost guarantee that anyone with a title like this is a cop.

Union rep

If your workplace has unionized, which I very much support, someone in your office (it could be you!) will be the union representative. They'll serve as the bridge between management and the workers. This is the person you can go to with grievances. This is the person you can go to when you get harassed by a manager, or a fellow worker. This is the person who's actually doing the job you were told human resources was responsible for. If your workplace isn't unionized, this person doesn't exist.

Look, just this morning I found out that the workers of the Medieval Times of Lyndhurst, NJ, had unionized. This is one of those dinner places where college students pretend to joust while you eat a thrice-reheated turkey leg, and the waiter asks if you'd like your goblet refilled with Sprite, milady. No shade, good for them. They didn't like their working conditions and they did something about it! It's a fairly easy process—it just takes a couple of folks at work deciding to call a union rep and setting up a conversation with them. You can do it, too.

Will management do shitty things if you decide to unionize? I'm not gonna lie, they might. But go back and read the section on management up there. They're already doing shitty things. Unionizing is self-care, boo!

You

You'll notice I enjoy beating a dead horse. But I'll say it again: every improvement in the workplace has been the result of workers banding together. And that usually starts with one person being fed up and brave enough to speak up in the hope that they aren't the only one. Trust that if your workplace needs improvement, there will be others who feel the same way. They may or may not be in a position to stick their neck out. They may not have the privilege to stick their neck out. They may not have the visa status to stick their neck out. But maybe you do.

PROGRESSING IN YOUR CAREER

I once had the pleasure of working with a very talented in-house designer at a large internet company. Let's call him Bob. Bob was a talented, good-natured guy. He knew his way around the company. He was always eager to help. And he had good ideas. I liked working with Bob.

One day, toward the end of the project, I'm having lunch (the swordfish) with Bob in the company's luxurious subsidized cafeteria. Bob says there's an opening for a design director and he's thinking about applying.

I encourage him to do so; after all, he's talented, good with people, and well respected within the company. He appreciates my encouragement and decides to apply.

A few weeks later we're back at our studio, having finished the project, and I get an email from Bob telling me he got the job. I congratulate him, excited that his confidence and hard work paid off.

About a year later, we got another contract with the same company. Unfortunately, we weren't working with Bob this time, but I sought him out. Bob looked like a different person. He had circles under his eyes, his skin was a whitish-green, and his studio was packed with piles of paperwork.

"How are you, Bob? Wanna grab lunch?"

Over lunch Bob described how his work life had turned into an endless chore of scheduling designers on projects, attending management meetings, approving vacations, and negotiating office politics.

"And I don't get to design anything anymore," he said.

Moving into management

Bob's story is sad, and not uncommon. I have way too many friends who've chased jobs because they were the next logical step in building a résumé. And obviously climbing the corporate ladder comes with a higher salary. We can't begrudge anyone the opportunity to make a better living.

But does the cost of making a better living have to mean giving up doing the thing you love doing? And how smart is a company that elevates people out of the thing they do best? There has to be a better way to reward people while leveraging their talent.

As your career advances, it makes sense to take on more and more responsibility. And to make more money for doing so. But do you want to be in charge of the work or of the people? Of all the people I know who've taken a job because it came with a "title," even while they knew they wouldn't be doing the kind of work they wanted to do, I can't think of one that looks back on that as a good decision. Not one.

You will be faced with the choice of how to take your career to the next level. Do you start your own company? Do you aim for design director at a large firm? Do you become a sought-after consultant where you only work six months of the year out of your swank modernist home in the hills?

The choice is yours. But I'd caution you to stay away from jobs that take you away from what you love to do, which is to design things. Although your definition of "designing things" may change. My friend John Gruber once said that Steve Jobs's greatest accomplishment wasn't designing any particular Apple product—it was designing Apple itself. You may get to the point where you're no longer designing specific products or specific websites, but instead helping to design the teams that design those things. And, eventually, designing the companies that those teams design within. But you don't stop designing.

If you're at a company where the next step up the ladder means managing people more than managing the quality of the design, get the hell out of there. There's way too much design to be done to be losing good people to absurd corporate structures that take our best designers out of commission. There are people who love managing people and scheduling their vacations. Leave that to them.

Directing others

Even the Greatest Designer Of all Time™ can only work so fast and do so much work. Eventually, either you'll be so successful that you'll have to figure out a way to increase your output without going nuts, or your company will decide that you're too much of a valued commodity to do the job of one person. Which means hiring other designers. Which means you're now directing them. The best way to get more value out of a designer who has reached a certain level is to have them teach what they know to other designers. (Just make sure you're in charge of their work quality, not their administrative overhead.)

I've been directing other designers at Mule for years. I absolutely sucked at it at first. (At least ten people just whispered, "You still suck at it.") It takes a while to get past your own competitive nature and inclination to just grab someone's mouse and say "Step aside!" Because, honestly, you probably can do what you're asking them to do a lot faster and most likely a lot better. But you'll be condemning yourself to a situation where you'll always have to jump in and be Batman, saving the city at the last minute from an inept police force, most likely made more inept because they know you'll swoop in when you're needed. (Ever wonder why Batman didn't just open up a secret facility and have Jim Gordon send him his best cops for some intensive training time? Because there's no such thing as a good cop! It was a trick question. 2022, y'all. Second edition showing no mercy!)

Leggo that ego

The first rule of directing others is that you have to let go of your ego. Or at least redirect it. The game is no longer about how well you can design something; it's about how well you can work with others so you can design well together. Otherwise, you'll be stuck in a situation where your capacity is limited to what you yourself can produce—not to mention being limited to ideas that only you can think of. The biggest joy for me in directing other designers is seeing them come up with ideas I never could have, and then working together to help them bring those ideas to fruition.

Give your people enough leeway to fail. And not so you can swoop in and save them, but specifically so they can trust that you won't. They need to learn to fall before they learn how to recover.

The designers you're directing need to be willing to show you work in progress. This isn't a presentation situation; this is intimate. This is sometimes looking-over-the-shoulder stuff. You know how much people like having someone hover over their shoulder? Not at all. Which means you need to establish trust. On both sides. Which means when you're looking over your designer's shoulder, you're the only one looking over their shoulder. Not you and the project manager, not you and the whole team. Just you. Your designers need to feel secure enough to tell you when they're stuck. And to trust you enough to help them get unstuck.

Setting expectations

People at different levels need to be directed differently. A younger, less experienced designer will need more frequent check-ins and more prescriptive feedback, while a more experienced designer may just need the occasional course correction.

But even more than that, a good design director will find out the best approach for each of the people they're responsible for. None of that works if people aren't willing to be directed, and if you aren't strong enough to do it. We've had designers come to work for us who'd either never been directed by anyone before, or had been damaged by bad design directors in the past. In both cases, you have to be clear and up-front about how your relationship is going to work. They need to understand when it's okay to push back, and when to acquiesce and just start doing what you tell them to, because that moment does come. Ultimately, you have a responsibility to deliver good work to your client. So training less experienced designers needs to be done within a setting where the client work doesn't suffer as a result. You can't deliver work with training wheels. The most dangerous instance in directing someone comes when they don't agree with you about their skill level. A young designer who believes they're better than they actually are won't be as open to your direction. They'll fight you and get defensive. Rather than being open-minded about the problem you're currently pointing out, they'll be wondering why you see a problem at all. They've closed themselves off to being directed, and that relationship isn't going to work.

Before we hire designers at Mule, I'm very clear with them about where I think they are talent- and experience-wise. I also tell them where I expect those skill levels to be in six months, and how we'll get them there. If they agree with the assessment and can commit to the work required to meet those future expectations, then we hire them. And with every designer I hire, I'm always looking for the one who's eventually going to take over my job.

BEING THE BOSS

Whether you're the lead designer on a project, or the director of a group of designers, or the head of a company, or the captain of a ship, your team's success depends on strong leadership.

I love bad TV. And for a while I was captivated by this ridiculous show called Whale Wars, where a ship full of well-meaning hippies and trust-fund babies getting back at their parents chased a Japanese whaling fleet around the South Seas. If ever a show had clearly defined "good guys" and "bad guys," this was it. Whales are awesome. No one wants to see a dead whale being pulled onto a ship and gutted. It's awful. And here were these well-meaning, if patchouli-stinking, kids primed to be the good guys.

But the captain of the hippie ship was a terribly inept leader. He was unable to make decisions. He made passive-aggressive comments to his crew. And ultimately he locked himself away in his quarters during the climactic episode of the season while telling his crew to handle the situation themselves. They were lost. They were looking for leadership that just wasn't going to come. And that's a horrible feeling. By the end of the series, I was actively rooting for the whalers to take their boat down. I am a terrible person, but the whales deserved a better champion.

Luckily, we don't work on ships and we're not bound by maritime law, so I can tell you that sometimes you need to just become the leader you're looking for without leaving yourself open to a mutiny charge.

Flat hierarchies are a lie

We're not complex creatures. We're chimps who know HTML5. And like chimps, we need to know someone is in charge. We get uneasy when the hierarchy is in flux. And we feel secure when our leadership is confidently taking us in the right direction. We like knowing what we're responsible for, and whether we're doing it well or doing it badly—especially if it also comes with instructions on how we can do it well instead. We like being rewarded in front of the other chimps, and being reprimanded behind the dignity of closed doors. We like to know whether we're doing our jobs well from the chimp, or person, who entrusted us with them.

There are people for whom leadership comes naturally. I'm not one of them. I started my own company because I wanted to be able to choose the types of jobs I worked on, because I was arrogant enough to think I could do it, and, honestly, because I didn't realize how hard it would be. And maybe that was a good thing.

When we first started hiring people, I was happy to let them find their way. Or, better said, I was too insecure in my own leadership abilities to point them in a particular direction. I was trying to be their buddy. Their fellow designer. They were looking for clear expectations on how to succeed. From their boss. And I failed them.

People want to do well. They need goals to meet. And you need to set those goals.

Hire people smarter than you

Everyone says this, but it's hard to do. It's much easier to hire people who are almost as smart as you, who are smarter than you at something you don't like doing, or (my all-time favorite) who remind you of yourself at a younger age. But to actually hire people smarter than you requires a heck of a lot of confidence and self-awareness.

Think of it this way, though: there are going to be people smarter than you out there. Where do you want them? Working for you, or for someone else? And surrounding yourself with smart, talented people is only going to make you better.

It's certainly the right call for the sake of the team.

And if you're honestly trying to build a team of incredibly smart people, be content that your place in the group doesn't have to be as the smartest one. You only need to be smart enough to convince them to work for you and keep them engaged.

Learn how to apologize

We've screwed up a lot over the years, but one particular screw-up stands out. We'd messed something up and the client was on the phone. He was pretty unhappy, verging on irate. Our project manager was attempting to placate him, explaining it wasn't actually that bad, that we'd fix it, that these things happen. She was doing her job well and offering him everything she was able to. Finally he asked to speak to me.

I picked up the phone. He furiously explained the problem. And I replied: "This is my responsibility. I apologize. It won't happen again. Now what can we do to fix it?"

Did we actually screw up? Doesn't matter. (But probably.) I realized that the only way we could get past his anger was to apologize. And that's not something I could ask someone else to do, nor would he have accepted it from someone else. As the boss, that was my sword to fall on.

You're going to screw up. Big. And when you do, you need to own it, clean it up, and move on. I've never held a mistake against an employee, but I can't tolerate someone who's unable to admit to their mistake. And when it's a client-facing mistake, you, as the boss, need to raise your hand and claim it.

Know when to part ways

If you think this section is about firing people, you're only partially right. That's coming up. But I recently said goodbye to one of my favorite designers I've ever worked with. She was talented, good-natured, smart, and a pure joy to have as a colleague. But she realized it was time to go do her own thing. And as sad as I was when she told me she was leaving, I was just as thrilled to hear she was confident enough to go at it on her own.

If you hire the right kind of people, they will eventually either leave you, or try to displace you. (My curse is not having been able to find the latter.)

Unfortunately, there are times when parting ways isn't such a happy event. Sometimes you hire the wrong people. Sometimes you hire good people and it turns out you just can't work together. There are lots of ways for relationships to go beyond fixing, but once they do, there's only one option. Being in charge means that, every once in a while, you're going to have to fire someone.

There are more pleasant names for it, of course: letting them go, telling them to start looking for other opportunities, laying them off. These are all bullshit. You're firing someone. Be honest and call what you're doing by its name.

I worked at a big company once. I was the designer in their pre-web marketing department. I'd been there for a while. I wasn't wild about recent changes in the company, I'd become kind of a prick to work with, and I was still young enough that I felt they owed me something. To make a long story short, I was done.

One day the president of the company comes into my office and tells me what a bad employee I'd become. (True.) He tells me I had two weeks to shape up or they'd fire me. I'd never been fired before! I hated the job, but I didn't want to be fired. I also needed the money. So I turned into a model employee for two weeks. (You can do anything for two weeks.) I mean, I did everything asked of me with a smile on my face. I honestly thought I'd averted being fired.

Two weeks later he comes into my office and fires me. I was devastated.

Hindsight is a wonderful thing. When I look back on that

moment, two things come to mind. First, I should have left that job a lot sooner. Second, I never had two weeks to shape up. The firing was done from the moment he gave me the warning. And I spent two weeks thinking I could save my job. (Actually, three things come to mind: I'd like to thank him for freeing up my time to learn web design.)

Because of this particular experience, I decided that I would never let an employee think they had a chance they actually didn't. When one of my employees is on thin ice, I let them know, and when I decide they're done, I let them know that.

I've had to fire a few people. I take no pleasure in it. But when someone's not doing the job, they're putting an undue burden on the other employees, especially at a small company, and it's not fair to them. It's also not fair to let someone continue in a job that you've decided they can't do.

Be clear in your communication with your employees. Give them regular updates on how they're doing. If they're not doing well, give them steps to correct the problem as well as a timeline for achieving those steps. Getting fired should never come as a surprise.

Being fired sucks. It sucks much more than doing the firing. So be clear and be quick and be humane. But make no mistake: it is their right to see you as the prick in that situation. And it is your responsibility to take the steps to make your company better.

Everyone I've ever fired has gone on to do well somewhere else.

LET'S WORK

I honestly don't think I've ever met a designer who didn't want to work. I've certainly met designers who were in situations where they felt like they couldn't do their best work, or situations where they were in over their heads. And I've met my share of knuckleheads who thought their work was better than it actually was. But we like to work. We define ourselves by it. To design is to work, and to call yourself a designer is to define yourself by the act of working.

Sometimes it takes us a while to find our groove, to find the place where we can do that work, but once we find it, we are happy to do it. Like every worker, we deserve workplaces where we are treated with respect and allowed to practice our craft.

Sometimes we need to create those places. The good news, my friends, is that creating those places is a design problem. And you are good at solving those!

CONCLUSION

I TOLD YOU A LIE AT THE BEGINNING OF THIS BOOK.

To be fair, I believed it when I said it. But I told you I was writing this book for you. I wasn't. I was writing it for me. Because this shit is hard.

Let's be honest. It's a pain in the ass to get up every morning and design stuff for people. Most of the time you're hitting your head against the wall, or going down the wrong path. Most of the work we do gets thrown out. Even the best of us have worse batting averages than ball players.

But those times when you get stuff right? Oh man, those are good times.

So I wrote this book to remind myself of those good times. To remind myself that the time I've spent filling up a digital trash can with metaphorical crumpled-up work—all the times I was asked to make logos bigger and buttons brighter, to "jazz things up"—actually amounted to something. It amounted to something I could pass on to you. And it paid back all the designers who were kind enough to help me learn the craft.

This thing we do is pretty great. We see a problem, we figure out how to fix it. Sometimes that problem is that no one knows where our band is playing, so we design a flyer. Sometimes no one knows where abortion services are available, so we design a website. Sometimes someone needs voting information translated into six languages, so we design a solution. Sometimes someone needs a database to track undocumented workers, and we don't design that.

Because where we put our labor is a choice.

Our labor is our consent.

For a long time, people have been trying to hire designers with promises that their companies are changing the world. It's a good line. So I'll ask you the same thing: Do you want to change the world? Because you can. Just not in the way those companies think.

We can change the world because we are workers. And we are many. The world isn't changed by billionaire assclowns in space; it's changed by workers with their feet on the ground. It's changed by workers who not only understand how to do the job, but want to ensure that their labor is being used to improve the world around them. The world is changed by workers looking out for one another and for the people of their community. The world is changed by workers who know how to use the tools to benefit others, and are willing to put down the tools to prevent others from being hurt.

To slightly paraphrase the great Margaret Mead: Never doubt that a small group of thoughtful, committed workers can change the world; indeed, it's the only thing that ever has.

You are all designers. You are all as good or as weak as you choose to be. And you all have an opportunity to leave your mark of good work on the world. Together, the mark you leave will be greater than any mark you can leave individually.

Goddamn.

THANK YOU

First of all, I'd like to acknowledge that this book was written on Ohlone land claimed by the US government in an unratified treaty. Land back.

Writing a book Is labor. And while there are a million types of labor harder than sitting at a nice desk, typing words on a bougie laptop, it can still kick your ass.

Revising a book is harder than writing a book from scratch. Like a true Philadelphian, my energy comes from proving other people wrong. I've written entire books to prove other people were being assholes (and then put their face on the cover). To revise a book you wrote ten years ago is to make yourself the asshole, and then spend months arguing with that asshole, before wiping him from existence.

I'd like to thank everyone who had to put up with me while I rewrote this book. I truly hope I am now less of an asshole than I was ten years ago.

I'd like to thank my editor Lisa Maria Marquis for turning my garbage words into something people will enjoy reading, and to tell me to add more—not less!—union stuff and jokes.

Thanks to Sam Cabrera for writing a kickass foreword.

Thanks to Heather Walls for proofreading.

We're still in a pandemic as I write this, so I'd like to thank everyone who was out there helping to keep the rest of us safe, fed, mobile, and alive. You've all had to put up with way too much shit for way too long. This debt, like many of America's debts, will never be adequately repaid.

I want to thank the whistleblowers, the union organizers, the walkouts, the rage quitters, the saboteurs, the leakers, the union card signers, the table flippers, the bolt cutter fetchers, and the people willing to call white men on their bullshit when they interrupt women during meetings.

I want to thank my wife Erika Hall for her amazing partnership.

I want to thank my daughter Chelsea for showing me what bravery looks like.

But mostly, I want to thank you. Thank you for reading this book. Thank you for giving a shit. Thank you for looking out for one another. Thank you for believing that we are capable of bigger and better things when we work together.

I love everybody who loves everybody.

Protect trans kids.

Make good trouble.

+++

APPENDICES

HELL YEAH, BONUS SHIT.

As a special thank you for buying the second edition of this book, I'm adding a few things just because I like you very much.

First off, A Designer's Code of Ethics, which, while it originated in my other book, *Ruined by Design*, is very much in line with the topic of this book. As always, I consider these a first take. Take 'em, make 'em better.

Secondly, there were two chapters that appeared in the first version of *Design Is a Job*, back in 2012, which were edited out of the second version. I stand by this decision. The second version is a much different animal than the first, and these two chapters break the flow. However, there's still valuable stuff in there for designers, so I'm tossing them in here as appendices because why not? It's my book, and you deserve it.

The first chapter is about how to present your work to people. Everyone has to do it. No one ever (rarely) gets any training on how to do it. So I hope this chapter helps.

The second chapter is, relatedly, about handling feedback. 'Cause after you present the work you're gonna get feedback on it. Hopefully.

Feedback is a gift, and if you work it right, you can get the kind of feedback that makes for better work. After all, you, like me, are a sack of rotting meat filled with biases. You cannot see what you cannot see. So getting feedback from other sacks of meat filled with different biases than you is amazing.

Thanks for reading. Now unionize, dammit.

APPENDIX 1

A DESIGNER'S CODE OF ETHICS

A few years ago, inspired by other professions' code of ethics, and surprised that I couldn't find one for designers, I took a stab at it. I gathered together a group of folks in the field and we came up with the list below. I consider it a living document. Make it better, keep improving it.

A DESIGNER IS, FIRST AND FOREMOST, A HUMAN BEING.

Before you are a designer, you are a human being. Like every other human being on the planet, you are part of the social contract. By choosing to be a designer, you are choosing to impact the people who come into contact with your work. You can either help them or hurt them with your actions. The effect of what you put into the fabric of society should always be a key consideration in your work.

We share a planet. Every human being on this planet is obligated to do our best to leave it in better shape than we found it. Designers don't get to opt out.

When you do work that needs income disparity or environmental damage to succeed, you are failing your job as a citizen, and therefore as a designer.

A DESIGNER IS RESPONSIBLE FOR THE WORK THEY PUT INTO THE WORLD.

Design is a discipline of action. You are responsible for what you put into the world. It has your name on it. And while it is impossible to predict exactly how your work may be used, it shouldn't be a surprise when work that is meant to hurt someone fulfills its mission.

We cannot be surprised when a gun we designed kills someone. We cannot be surprised when a database we designed to catalog immigrants gets those immigrants deported. When we knowingly produce work that is intended to harm, we are abdicating our responsibility. When we ignorantly produce work that harms others because we didn't consider the full ramifications of that work, we are doubly guilty.

The work you bring into the world is your legacy. It will outlive you. And it will speak for you.

A DESIGNER VALUES IMPACT OVER FORM.

We need to fear the consequences of our work more than we love the cleverness of our ideas.

Design does not exist in a vacuum. Society is the biggest system we can impact, and everything you do, both good and bad, is a part of that system. Ultimately we must judge the value of our work based on that impact, rather than on aesthetic considerations. An object that is designed to harm people cannot be said to be well designed, no matter how aesthetically pleasing it might be, because to design it well is to design it to harm. Nothing a totalitarian regime designs is well designed, because it has been designed by a totalitarian regime.

A broken gun is better designed than a working gun.

A DESIGNER OWES THE PEOPLE WHO HIRE THEM NOT JUST THEIR LABOR, BUT THEIR COUNSEL.

When you are hired to design something, you are hired for your expertise. Your job is not just to produce that work, but to evaluate the impact of that work, and to relay that impact to your client or employer. Should that impact be negative, it is your job to help your client or employer, if possible, to eliminate the negative impact of the work. If it's impossible to eliminate the negative impact of the work, it's your job to stop it from seeing the light of day.

In other words, you're not hired just to dig a ditch, but to evaluate the economic, sociological, and ecological impact of that ditch. If the ditch fails those tests, it's your job to destroy the shovels.

A designer uses their expertise in the service of others without being a servant. Saying no is a design skill. Asking why is a design skill. Rolling your eyes is not. Asking ourselves why we are making something is an infinitely better question than asking ourselves whether we can make it.

A DESIGNER WELCOMES CRITICISM.

No code of ethics should protect your work from criticism, be it from clients, the public, or other designers. Instead, you should encourage criticism in order to create better work in the future, and be open to that criticism coming from anywhere. If your work is so fragile that it can't withstand criticism, it shouldn't exist. The time to kick the tires on your work comes before those tires hit the road.

The role of criticism, when given appropriately, is to evaluate and improve work. Criticism is a gift. It makes good work better. It keeps bad work from seeing the light of day.

Criticism should be asked for and welcomed at every step of the design process. You can't fix a cake once it has been baked. But you can increase the chances your project is successful by getting feedback early and often. It's your responsibility to ask for criticism.

A DESIGNER STRIVES TO KNOW THEIR AUDIENCE.

Design is the intentional solution to a problem within a

set of constraints. To know whether you are properly solving the problem, you need to meet the people who have it. And if you are part of a team, your team should strive to reflect those people. The more a team can reflect the audience it is solving for, the more thoroughly it can solve the problem. That team should come at the problem from different points of view, from different backgrounds, from different sets of needs and experiences. A team with a single point of view will never understand the constraints they need to design for as well as a team with multiple points of view.

A DESIGNER DOES NOT BELIEVE IN EDGE CASES.

When you decide who you're designing for, you're making an implicit statement about who you're not designing for. For years we referred to people who weren't crucial to our products' success as "edge cases." We were marginalizing people.

We were making a decision that there were people in the world whose problems weren't worth solving.

Most products would consider their edge cases to be only one percent of their users. Facebook now claims to have two billion users. One percent of two billion people is twenty million people. Those are the people being marginalized.

Trans people get caught on the edges of "real names" projects. Single parents get caught on the edges of "both parents must sign" permission slips. Elderly immigrants get caught on the edges of voting when they can't get ballots in their native tongues.

There are no edge cases. There are only human beings, and we owe them our best work.

A DESIGNER IS PART OF A PROFESSIONAL COMMUNITY.

You are part of a professional community and the way you do your job and handle yourself professionally affects everyone in that community. Just as a rising tide affects all boats, taking a shit in the pool affects all swimmers. If you are dishonest with a client or employer, the designer behind you will pay the price. If you work for free, the designer behind you will be expected to do the same. If you do not hold your ground on doing bad work, the designer behind you will have to work twice as hard to make up for it.

While a designer has an ethical obligation to earn a living to the best of their abilities and opportunities, doing it at the expense of others who share the craft is a disservice to us all. Never throw another designer under the bus to advance your own agenda. This includes public redesigns of someone else's work, spec work, unsolicited work, and plagiarism.

A DESIGNER WELCOMES A DIVERSE AND COMPETITIVE FIELD.

Throughout their entire career, a designer seeks to learn. That means confronting what they do not know. That means listening to other people's experiences. That means welcoming and encouraging people who come from diverse backgrounds and cultures. That means making space at the table for people who society has historically kept down. We must make space for traditionally marginalized voices to be heard in the profession. Diversity leads to better outcomes and solutions. Diversity leads to better design.

A designer keeps their ego in check, knows when to shut up and listen, is aware of their own biases and welcomes having them checked, and fights to make room for those who have been silenced.

A DESIGNER TAKES TIME FOR SELF-REFLECTION.

No one wakes up one day deciding to throw their ethics out the window. It happens slowly, one slippery slope at a time. It's a series of small decisions that might even seem fine at the time, and before you know it, you're designing filtering UI for Walmart's online gun shop.

Take the time for regular self-reflection. Evaluate the decisions you've made recently. Are you staying true to who you are? Or are you slowly moving your ethical goal posts a few yards at a time with each raise or stock option award?

Have you veered off course? Correct it. Is your workplace an unethical hellmouth? Find another one.

Your job is a choice. Please do it right.

APPENDIX 2

PRESENTING WORK

A DESIGNER WHO DOES NOT PRESENT THEIR OWN WORK ISN'T DOING THEIR FULL JOB.

Presenting the work, explaining the rationale, answering questions, and eliciting feedback are part of the design toolkit. If you sit at your desk while someone else presents work to the client you don't get to complain about the feedback. The failure was yours.

I get it. It's nerve-wrecking. And I'm guessing most people reading this went through design programs where nobody bothered to teach you how to present your work. Critiques don't count, everyone in the room is trained about the same stuff that you are, they all care about design, and are *thrilled* to get a thirty minute presentation about typography. This is not a real-world scenario.

I've been presenting design to clients and internal teams for a long time (Bush, Sr., was in the White House when I started), and I still get anxious about it. It may have been a while since I've thrown up in a client's bathroom and washed off next to the person I'd eventually be presenting to, sure, but I still get nerves.

The fact that I'm sitting here telling you how to present design isn't a testament that I was born with some great skill—it's a testament that anyone can present design if they're actually taught how to do it, and willing to pick themselves up off the floor and keep at it.

There is no substitute for learning how to present design other than presenting design. Be nervous. No one ever died from a poor presentation. At some point you'll realize that your nerves are lying to you. The evidence that you can present design will grow too large to ignore. You'll do it again and again, and one day you'll realize that you're not as terrified as you used to be. Stay there. You're actually designing from that point forward.

WHERE NERVES COME FROM

It's worth discussing why people get nervous when they present, if only to decouple it from the reasons people believe they get nervous when they present. Get in, we're doing science.

The one and only reason people get nervous when they present work is because they are designed that way. Your amygdala, which is the part of your brain in charge of keeping you alive, senses a presentation as a threat. Quite simply, there are more eyeballs staring at you than you have staring back at them. This triggers a flight or flight response which releases 500 metric fucktons (actual science number) of adrenaline into your system so you can either run away from or fight these people off. Our historical ancestors found this very helpful, you—standing in front of a room of peers, and the leadership team, and that jerk Chet from Marketing?—not so much. But the real villain in our story is imposter syndrome, which thankfully we got rid of in a previous chapter. But imposter syndrome will start whispering things in your ear like "you are nervous because your work is terrible" or "you are nervous because you are being judged" or "you are nervous because you're about to waste everyone's time." And while some of those things *may* be true (this chapter will try to minimize those possibilities), the one and only true reason that you are nervous is because your body is now coursing with adrenaline. That's it. Imposter syndrome, which I will keep reminding you we got rid of in a previous chapter (go re-read it if you need to) is hijacking a natural, if now misbegotten, process to fuck with your head.

So the next time you are about to present work, know that you will get nervous, remember that it's a sign that your body and mind are working as they should, name it, put it aside, and do your thing.

Science!

IT'S NOT GOING TO SELL ITSELF

The biggest myth ever perpetuated in the design field is that good design sells itself. Design can't speak for itself any more than a tamale can take off its own husk. You're presenting a solution to a business problem, and you're presenting it as an advocate for the end users. The audience needs to know that you've studied the problem, understood its complexities, and that you're working from that understanding.

Stop trying to get your audience to "understand design" and instead show them that you understand what they hired you to do. My plumber doesn't need me to "understand plumbing" in order to fix my sink, they just need to convince me they fixed it. Explain how the choices you've made lead to a successful project. This isn't magic, it's math. Show your work. Don't hope someone "gets it," and don't blame them if they don't—convince them. And if you can't convince them, it might be worth listening to why!

If you're confident enough that you're putting the right solution in front of someone, which you better be, you need to pass that confidence along to them. Ultimately, your job is to make other people feel confident in the design. Confidence is as much of a deliverable as anything you're handing over in the project.

WHY YOU NEED TO PRESENT YOUR OWN WORK

It bears repeating.

Being able to present your own work is a core design skill. It helps build rapport with the audience. It puts the person directly responsible for the work in front of them. It shows them that you're presenting that work with confidence. And it gives them an opportunity to ask questions directly of the person who did the work.

When I was a kid we played a game called telephone. All the kids form a circle, and the lead kid whispers a word in the ear of the kid next to them. That kid whispers the word to the kid next to him, and so on, until the word gets whispered in your ear. (Shut up, we didn't have video games or the internet.) Without fail, the word that got whispered in your ear was different than the word you started with. "Chocolate" would become "chunky latte," "baseball" would become "base board," and "I need you to work this weekend" would become "you need to unionize the workplace." You get the idea.

When clients are speaking to you through an intermediary you're not hearing directly from them. You can't see their face. That's important. Quite often the look on their face will tell you much more than the words out of their mouth.

Things that were posed as questions, such as, "Why is the logo that size?" get reported back to you as statements like, "Make the logo bigger." You weren't there to answer the question. In fact, you didn't even know it was posed as a question.

As a designer, you need to become adept at interpreting what a client actually means. You need to be there to ask clarifying questions. You need to manage the presentation process as carefully as you make the work.

This also applies to work that's not being presented in person. If you're working with a long-distance client, set up a video call so that you can present the work and the rationale. And don't send the work in advance. Present it during the call, after you've had the opportunity to set the stage properly. The last thing you want is clients getting ahead of you and starting a call with their list of proposed changes before you've had a chance to walk them through your rationale.

If you're working at a company that won't let you present your own work, make the case for how it benefits everyone. It helps you, it helps the team, it helps the project get done correctly. It saves time, which ultimately saves money. If they still won't let you present your own work, get the hell out. They're not letting you do your job. Don't work at a place that doesn't let you do your job correctly.

HAVE AN AGENDA

Before you ever get in a room with anyone, let them know why you're getting in that room. (If you're not sure why you're meeting, then maybe there's no reason to meet.) Make sure everyone is walking in with the same expectations and knows what's expected of them. Also, let them know how long the presentation is scheduled for, and give them the desired outcome.

A meeting without an agenda is an invitation for everyone to bring their pet issues. After all, you can't really stray off course if no course has been set.

BE CONFIDENT IN WHAT YOU'RE SHOWING

There should be absolutely no doubt that this is your meeting. Stand up. Walk to the head of the room. Announce who you are, what you do, reiterate the goals of the project and tell people how what you're about to show them will meet those goals.

Put your audience at ease by letting them know that they're exactly the right people to have in the room, and what you need from them is their expertise with the product or service, not design knowledge.

Steer the discussion away from personal subjectivity by outlining good topics of feedback: how well the proposed solutions meet specific metrics, whether their voice and brand are coming through, etc. Encourage them to stay in their own zone of expertise and they won't attempt to hop on yours.

Never apologize for what you're not showing. By the time you're presenting, you should be focused on presenting what you have, not making excuses for what you don't. And you need to believe what you're saying to convince the client of the same. If you think the work is on the way to meeting their goals then say that. Design is an iterative process, done with a client's proper involvement at key points. The goal isn't always to present finished work; it's to present work at the right time.

I've met a few designers over the years who feel like selling design is manipulation. Manipulation is convincing someone that the truth is different than what it seems. You're familiar with the marketing phrase, "Sell the sizzle, not the steak"? Well, if the steak sucks, that's manipulation. If the steak backs up the sizzle's promise then it's good salesmanship. If you believe the work is good, and you can stand before a client and make an honest attempt at convincing them of the same by using the research you've done, then what you are doing is selling good design.

Don't be arrogant. Confidence means you believe you'll succeed because you've done the research, you understand the problem, and the work you're presenting backs that up. Arrogance is believing you'll succeed despite not having done those things. Don't be that jerk.

SPEAK TO GOALS, NOT FEATURES

Your review of the work should be goal-driven. And prioritized. The client had major goals with this project. They'll be anxious to see how those goals are addressed. Go there first, and do it in a way that maps to their goal and not to your manifestation of the goal. For example:

"Here's a Twitter button!" is bad.

"As you can see, we've addressed your social media concerns!" is better.

"We've placed social media sharing tools in places alongside the article where testing shows the reader is most likely to interact with them, and designed that space so that it's flexible enough to add and subtract services as they rise and fall from prominence," is the best of all.

Remember that every feature is there for a reason. Lead with the reason. And if you don't have a good reason why it's there, that's probably a sign it's not needed.

DROP A BODY ON THE FLOOR IMMEDIATELY

I am a fan of mystery novels, and maybe you are too. A good presentation should start like a good mystery novel: drop a body on the floor as soon as possible. Show people the thing they need to see and then give us the story about how it came to be there.

Most presentations follow a chronological timeline of the designer's effort. First I did this, then I did this, here we are writing words on post-it notes of different colors, here's some of the directions we eventually abandoned, etc, etc. It's an exhaustive comprehensive attempt to be graded on effort. We want to show the room how hard we worked, which puts *us* at the center of the presentation, and we should never be at the center of a presentation, it should always be about the work. This is, however, how most people learned how to present if you equate presentations with critiques. Everyone in the room wanted to know *how* we did it, mostly so they could steal it.

Your audience wants the work, they're itching for it. If you've ever looked out at your audience and seen them checking their phones, or worst yet, telling you to get on with it, it's because you've exhausted their patience. Drop a body on the floor. Then, like a good mystery novel, explain how the body came to be. Now that their expectations have been met they're ready to hear the story. Talk about how the research led you here, talk about how the strategy evolved. Now they feel like they're meeting the chef *after* having eaten a good meal, not being dragged through the kitchen on the way to their table. Yes, I'm mixing metaphors. Shut up.

DON'T GIVE A REAL ESTATE TOUR

The most obvious sign that a designer is nervous is the real estate tour. You've seen this. The presenter will start at the top left of the page, telling you where the logo is, describing how it "pops," then move right and down, describing every element along the way, taking you on a guided tour of every element on the page, winding up at the copyright notice in the footer. They'll make no mention of goals or benefits. In the process they'll leave themselves totally open to clients pulling up with construction teams and cranes to do some remodeling.

Don't waste a client's time walking them through what they can already see. Your job is to explain how what they're looking at is the best way to achieve their goals.

If you start with the declaration that you're accommodating the social media strategy, there's a lot less resistance because you'll be showing that every visual element is in a particular place for a reason. Otherwise it just sounds like you're dropping buttons in random spots.

GIVE THE CLIENT PERMISSION TO GO NEGATIVE

It can be incredibly difficult for clients to say anything negative about your work. Something about "upsetting the sensitive, talented creatives." Or just basic, if misguided, decency. This doesn't help the project. Unspoken expectations lead to seething unspoken frustration, which ultimately bursts forth in a gruesome mess when you've run out of budget.

Let them know that negative feedback is an essential part of the process and a part of the client's job. We prefer constructive and well-reasoned negative feedback, but even an outburst of "Never show me that crap again!" saves everyone a lot of time.

At the beginning of every visual design presentation, particularly in the early stages, we give a little speech, some variation of, "Today, we are going to show you some things that may not be right. If you see something that isn't working, you need to point it out. If you don't tell us what you think isn't working, we will show you the same thing again and again until we are out of time and money and you are stuck with it."

We've found this exercise makes everyone a lot more comfortable with the process and saves a lot of time wasted by clients saying, "Yeah, that seems fine" when they're thinking, "That is a pile of noxious crap for our customers to poop on. I hope they can read my mind so I don't have to say it."

HELP PEOPLE GIVE YOU THE INFORMATION YOU NEED

More often than not the designers who complain that clients give them subjective feedback are the ones most prone to asking for subjective feedback. Everyone gets the feedback they deserve.

Stay away from questions like "Do you like it?" You weren't hired to make something they liked. Ask your clients to comment specifically about the things you want them to address. Avoid talking about things you don't want them to go into. "What do you guys think of the size of the logo?" for example.

They will, of course, still give you feedback about things you may not want feedback about. Most of it you just need to hear and acknowledge. But if they're persistent, ask them to map what they are asking for to a goal, and follow up with clarifying questions. Oftentimes a client's prescriptive feedback needs to be unpacked to get to the real issue.

The best way to get helpful feedback is to be as specific as possible with the person you are asking, for example: "Tim, as the lead engineer, do you see anything that might trip your team up or cause you to miss your deadline?" gets you much better feedback than "Tim, what do you think?"

WHEN PRESENTING WITH A TEAM

If you're presenting with a team, you and everyone else has a role to play. Your job is to take the responsibility for selling the design work. But you may not be the only designer on the project. If you've been working collaboratively, present collaboratively. Just make sure that you've clearly outlined who's taking charge where.

You may also have a project manager on your side who does the initial project recap. They should also keep track of time management and the agenda.

And while everyone should have a set role, be flexible enough to let everyone play to their strengths. If your interaction designer has been championing the faceted navigation model you're proposing, let her pitch it. Weave other people into the presentation as seamlessly as possible.

Always present with a united front. No matter how much arguing there might have been back at the office, by the time you're in front of a client, every solution you're presenting is a team solution. Should a client propose something similar to what had been shot down internally earlier in the week, anyone who screams, "That's what I wanted!" is throwing the rest of the team under the bus. Graceless.

BE OPEN TO BEING WRONG

Confidence also means you can handle being wrong. And you will be. Many times. The goal of a presentation isn't to shove your decisions down a client's throat—it's to present what you believe to be right in a manner that also persuades others that it's right. And it also means allowing them to convince you that you're wrong. I can't tell you how many times I've walked into a presentation totally confident that what I was presenting was rock solid, only to have a client come up with an obvious use case that I'd overlooked and which brought the whole thing to a crashing halt. And that's good. Because part of the design process is finding those holes. Be glad you're working with a smart client. We are all rotting sacks of flesh with our own biases, part of the reason we show work to other people is to have those biases exposed. Don't keep trying to sell something that's broken. Acknowledge you were wrong, and be thankful for the new insight.

END STRONG

End on time. Respect your client's time and show them that yours is valuable as well.

Make sure everyone's questions have been answered. And make sure they have an avenue for feedback, which we'll discuss in the next chapter, as well as a deadline. Thank them for their time. Shake hands and offer to take your glass to the sink. They'll tell you to leave it, but offer anyway. Your mother raised you well.

Which reminds me of a fabulous story from the pre-Mule days. I was working with another studio and we'd just gone through a particularly difficult presentation. Couldn't wait to get out of the client's office. We said our goodbyes, packed up our stuff, and headed to the elevator. While we're waiting for the elevator our creative director turns to me and says, "God, they were assholes!" just as one of said assholes came walking around the corner.

The rest of the project was awkward.

My advice to you when you finish the presentation is to thank everyone for their time. Shut your mouth. Pack up your laptop. Head out of the building. Walk out the front door. Around the corner. In the car. Across the bridge. Okay, now you can talk.

Or, just make it a habit never to speak ill of your clients. They're paying your bills. And putting their livelihood in your hands. They're good people. And if they're *not* good people? Why the hell did you sell them your labor?

BUT FOR REAL, IGNORE THIS WHOLE THING

I'm kinda kidding, and also kinda not. The best thing I can tell you about presentations, and maybe I shoulda led with this, is that they don't exist. That's great news if presentations make you nervous. Because, honestly, most people love talking about their work. We love doing it and we can't shut up about it. I know this for a fact because I've invited designers into my home to do things like watch basketball, and from the moment they get there to the moment they leave, they talk about their work. It's exhausting. Designers can talk about their work for hours. And yet, when they're told they need to present something they freak out, they start making slides, they write scripts, they work themselves into a lather. Eventually they go from being someone who's been working on something for the last eight months, to someone who's been putting a presentation together for the last couple of days.

The person I want to talk to, though, is the person who's been working on solving the problem that was handed to them for the last eight months. I want the person who remembers every nook and cranny of the problem, not the person who remembers the script they wrote two days ago. I want to have a conversation with the expert I hired to solve the problem, not see a slideshow that was put together to impress me.

You've been preparing for this presentation conversation the entire time you've been on this project. Talk about what you know. So sure, take some time to gather your thoughts, come up with a good story, and maybe even toss a few slides together, but show up as the person who's doing the work, not the person who's doing the presentation, and you'll be fine.

APPENDIX 3

MANAGING FEEDBACK

GATHERING AND MANAGING FEEDBACK IS AN ESSENTIAL PART OF THE DESIGN PROCESS.

And next to Adobe updater updates, it's the thing designers complain about the most. Complaints range from "They're telling me what to do!" to "They're *not* telling me what to do!" to "Oh, God! They sent a PowerPoint comp!"

We're far enough along in this book now that I bet you can guess what's coming. That's right: I love you, but it's your fault. You can't assume clients will know how to give you the right feedback at the right time and in the right manner. What you can hopefully be confident about is that both you and the client are trying to get to the same goal: a successful project. So take a deep breath. And let's figure out how to get what you need.

SETTING THE STAGE FOR GOOD FEEDBACK

No matter how smart or articulate a person is, no one intuitively knows how to give useful design feedback. It's a learned skill. And it's a skill that's not part of anyone's day job (except for art directors, and they're exhausting). How to give feedback must be taught. And the teaching starts at the very beginning of the project. When you decide to take on a project, and explain how the project will go, make sure you explain how their feedback is crucial to making the process work. Both of you have a role to play and the feedback is the rich sinewy connective tissue between both parties.

Has your partner worked with designers before? How often, how long ago? And what role did they play? Get a proper gauge of their experience with giving feedback on a design project. And realize that "A lot!" may not be a good answer. Experience working with freelancers may not translate well to working with an in-house team, and vice versa. And neither translates to working with a small shop. In the end, you will have to both show your partner how to work with you and, more importantly, you'll have to learn how to work with your partner.

There's nothing less helpful to the client than telling them you want open-ended "feedback." Your clients need your help in figuring out what kind of feedback you're looking for. They need structure and guidance. And if you provide that structure and guidance ahead of time there's a far greater chance that you'll be getting the kind of structured feedback you'll need to move forward, and save everyone a lot of time.

Work with your client to establish a feedback cycle that works for both of you. At Mule, we generally like getting feedback no more than forty-eight hours after a presentation so we don't lose momentum. Some companies have too many stakeholders to make this possible, so neither should you require it, nor should they agree to it in that case. You'd both be setting yourselves up to fail. Come to a good compromise. But make sure they understand that more time for feedback means the project completion date moves back as well.

Some companies move faster, and want to promise you feedback in twenty-four hours. That's great! Give them forty-eight. Companies that move faster also have to put out a lot more fires, and generally with a smaller staff. It's probably the people who are supposed to be giving you their feedback that are putting out those fires.

Sometimes companies are moving so fast, that your best bet is to get the feedback you need then and there during a presentation. It might not be as thorough, or well-considered, but it's sometimes more honest, and it means you can get right back to work, which I'll take in a heartbeat.

Either way, document the agreed-upon feedback cycle in writing and put it in the Statement of Work. It's a deadline, subject to the typical accolades that come from meeting a deadline and remonstrations that come with missing one.

THE INVISIBLE STAKEHOLDER

This is also when those invisible stakeholders you didn't do a good enough job of ferreting out make their presence felt. We've all met these lovely people. They show up during the third week of development with thoughts about the design concept. Or the team you've been working with for three months will announce that once the site is ready to "launch" they'll unveil it to the board to get their thoughts. You will feel tricked when this happens. You know how you should really be feeling? That's right—irritated with yourself for not flushing them out early in the process.

Going into a project, you need to know who on the partner side provides input, who gives feedback, and who approves. You may have a better idea of who these people should be than your client. As you meet people within the organization during the discovery phase (because you have one, right?) ask yourself whether they're someone who can insert themselves into the feedback cycle later. Be proactive and have a discussion with your main contact about involving them sooner rather than later. If they decide they won't be participating in reviews, then you'll have to hold them to that later. As important as it is to have a small feedback group reviewing design decisions, it's even more important that it's the right group.

The general rule is to involve as many people as possible

in early discussions, and make that number as small as possible once you go into review cycles. Often, people on the client side say they want to remain involved, but really just want to tell you the one thing they care about and then get back to their day job. If they don't get to tell you that one thing, then they will pop up at the end of the project all bent out of shape.

YOU DON'T WANT DESIGN FEEDBACK

Most clients tend to feel self-conscious about voicing their opinion about design. How often have you heard "I don't know anything about design!" That's fine; they don't have to. What they do know something about, however, is their business, and what you're actually looking for here is business feedback. You need to know whether the design work is meeting the business goals of the project. You say "design," and clients hear "art," and now everyone's uncomfortable. Act like the design expert the client hired, remind them—if need be—that they hired a design expert, and their job is to be the business experts. Keep them in their comfort zone and they're less likely to come back with feedback from a family member who's got "a good sense of design."

So rather than asking the lead developer if they have "any feedback" or "design feedback" try asking them if they "see anything that might cause any development issues for their team." Ask the branding team if they see anything that feels out of place from a branding perspective. Ask the research team how well this work reflects the research they did. Etc. Get specific, and make it about their expertise. Keep them in their comfort zone because that's the information you actually need to improve the work. Getting the lead developer's feedback on design might validate you, but getting their feedback on how it affects the development cycle validates, and might possibly improve, the work.

Of course our dirty little secret is that while we're telling our clients to be objective, and to keep personal preferences out of it, there's nothing we want more than them to like it. The first step in getting objective feedback is to make sure we're giving objective presentations and keeping our own feelings out of the mix.

FEEDBACK IS NOT ABOUT YOU

Too often we use feedback for our own validation. We talk about positive feedback, and define it as "stuff that makes us feel good," and we talk about negative feedback, and define it as "stuff that makes us feel bad." But that's putting ourselves at the center of the conversation. I believe this happens because most designers are still mapping presentations to the kind of design critiques they did in school. You do some work, you present it to the class, tell them about all the design challenges you faced, everyone nods along because they all faced the same challenges with their projects, you make sure the teacher knows how hard you worked, and then everyone gives you feedback based on how much they like you, and you get a grade. It's all very validating. It makes sense that a school critique is about you, after all *you* are paying a lot of money to be there. (You're still probably paying it off.)

But in a professional setting, where people are paying *you* to solve a problem, the tables turn a little bit. The goal of real-world feedback is to make the work better, not to get a grade. There is no positive or negative feedback. There is useful feedback and useless feedback. And everyone in that room is, hopefully, trying to break your work.

Walking out of a presentation with a long list of things to fix is exactly what you want, and making sure that people are willing to give you that list is exactly what you should be aiming for.

FEEDBACK GUIDELINES

We prepare feedback guidelines for our design partners before presentations. We tell them what's helpful to talk about and what's not. They're thankful for the guidance and it saves them a lot of time. People love it when you save them time, and they love it when you give them explicit instructions on how to succeed at something. We tell them what decisions are relevant at this point, what to ignore, and what may require further clarification. We also reiterate key points made during the presentation so they don't forget.

For example, let's say we just did our first visual concept presentation for a website. If we did it right, then the comps we showed were loose and sketchy and focused on tone, not specific functionality. And they were probably done before site architecture was finalized because one informs the other and everyone's working fast. The last thing we want is a lot of feedback about missing functionality. So the feedback guidelines might say:

Focus on:

- Overall tone (Does this site feel like your company?)
- Voice (Is it using language the right way? Is it friendly? Authoritative? Trustworthy?)
- Structure (Too dense? Not dense enough?)

Ignore:

- Specific labels and body copy (Why is everything in Latin?)
- Missing elements (Where's my newsletter signup?)
- Things we did not show you (I have comments about the company history page)

It's also a good idea to address personal preferences before any feedback is given, rather than after, at which point you risk making people feel dumb for liking green, which although not the right color for this project, is a very lovely color in and of itself. Unlike mauve. And if someone tells you they like green or don't like green just thank and move on. If you start arguing about the subjective stuff you're giving it air. You're making it real. But you'll be amazed how a well-placed, polite "Thank you!" puts so many non-issues to bed.

Feedback guidelines should be short enough to be helpful, and not long enough to be considered additional work. Go over them with the client during the presentation and make yourself available should they require more clarification. Make sure every member of the team gets them so they're all playing under the same rules.

And know that they'll give you feedback on everything you asked them to ignore anyway. Clients will gravitate toward what they're worried about the most, or what they're most comfortable with. Feel free to ignore it until the time is right.

THE MOST IMPORTANT FEEDBACK GUIDELINE OF ALL

Many clients believe that proper feedback means telling a designer how to fix something, or what is known as prescriptive feedback. Many designers complain about prescriptive feedback, unless they don't get it. At which point they complain they don't know what the client wants. Yes, our reputation as being difficult is well-deserved.

Let's remember that clients aren't experts in giving design feedback and—I'm gonna go out on a limb here—venture that most of the time they're the problem solver on their teams. Let them know that not only are these your problems to solve, but that they're paying you to solve them.

You need to start holding this line during presentations, as clients jump in with suggestions on things they're not happy about. Impromptu prescriptive work sessions during presentations can eat up a lot of time, and put you in the awkward spot of having to tell clients that their suggestions may not be very good. So let them know ahead of time, "If something isn't working, tell us why and we'll fix it. We want to earn our money." Remind them again in the written guidelines. (Have I turned presentations into impromptu working sessions? Yes, and sometimes with great results. But you really, really, really, need to trust and be comfortable with everyone in the room.)

You don't want to put yourself in a position of carrying out your client's design decisions. If you allow them to start designing for you, you have no one to blame but yourself. If they insist on prescriptive feedback it's time for an account-level conversation. Anyone who hires you because they thought you could do the job and then doesn't let you do it has lost respect for either you or the design process.

Can you put "no prescriptive feedback" in the contract? I suppose you could, if you really wanted to. But ultimately I think this is one of those times you just need to be able to work without a net. Keep the contract to legal issues, and use your design skills and backbone to support your design process.

GETTING FEEDBACK ON TIME

As I mention above, set up a deadline for turning feedback around. Any time you're requesting feedback, remind the client how much time they have and get them to commit to meeting the deadline. (The success of every project is the hundreds of little commitments made along the way.) Let them know that if something comes up that might put the deadline in jeopardy, they should let you know right away. That goes for both parties. The minute a deadline is in jeopardy notify the other party. There are many reasons for missing a deadline, but there is only one for not notifying the other party. (That's a free life tip, kids.)

The longer clients are taking to get you feedback, the more likely that they're not just pointing out issues, but working out solutions. You, or your project manager, need to get them on the phone. Ask them if they need your team's help with anything. Remind them to come back with questions, and not solutions. That usually goes a little something like this:

"What's up with our feedback, Bob?"

"We're not happy with global nav, but we're not sure how to fix it."

"Well, just tell us why it's not working for you. We'll come up with something. That's why you're paying us, Bob."

"That'd be swell. Cause we're stumped."

Most clients try to fix things because they believe it's the right thing to do. And unless you tell them it's not you have no one to blame but yourself.

HOW TO READ CLIENT FEEDBACK

Eventually the blessed day will come when the client sends you their feedback. Oh, did I tell you to get it in writing? You need to get it in writing. An email is just fine. Immediately acknowledge getting it, thank them for sending it. And sit back. Relax. It's important that you're relaxed when reading client feedback. Make some tea, or take a CBD bath. Open your mind to the possibilities that you might be going on a journey. Client feedback has a way of taking you to places you never dreamed you'd go. But, honestly, remember two very important things:

You're both working toward the same goal with the best of intentions.

Most clients haven't been trained in how to give feedback. (Well, except by you.)

First off, do yourself a favor and just read the whole thing straight through at least once. Resist the urge to read it aloud to the rest of your team, if you have one, because mocking a someone's feedback reflects poorly on you. They're gonna say some weird stuff. They're honestly trying. Approach it with grace. Don't share the feedback with anyone else until you're sure you can do it in a calm, reassuring manner. (Pro tip: sometimes my partner Erika withholds feedback from me for days. We're totally in a "do as I say, not as I do" zone here.)

Your job is now to take the client's feedback and put it into an actionable form.

ORGANIZING FEEDBACK

Your first job is to separate the actionable feedback from the non-actionable feedback. Sometimes clients just like to document their thought process. Your job is to sift through and find the actionable from the non-actionable.

"At first I thought this work reminded me of something I didn't like, but then it didn't and then it did again and then the reason I thought it did was because the typeface was similar, but then I changed my mind again and realized it was because it was blue."

There is absolutely nothing you can do. Let it go.

On the other hand, "We've changed our strategy!" is not feedback. Nor is anything that even hints at it. It's a shift in goals to be addressed immediately at the account level.

Having separated the feedback from the non-feedback, it's time to make three lists.

- ***To-do***—No brainers. Stuff you can change immediately.
- ***Disagreements***—Things that run counter to the goals of the project, disrupt the design, break conventions, or don't jive with the client's brand. Or, they're just bad ideas. Don't do bad design! Prepare to make your case to the client about every issue on this list. You're going to have to convince them. It's usually not as hard as you think. Good clients listen to reason more than you give them credit for. You're just going to have to persuade them. Part of the job.
- ***Clarification needed***—If you don't understand something on the client's feedback list, pick up the phone and ask them about it. Don't try to read their minds. Once you've gotten clarification on this list, move to one of the other two lists.

If you have a team now is a good time to go over the list with them. You'll be organized and relaxed, and seeing that you've called out issues to discuss further with the client will probably keep your team relaxed as well.

Let's get on the phone.

GOING OVER FEEDBACK TOGETHER

Have we talked about how awesome talking to your client in person is? Every time you deal with your client in person is an opportunity to build trust. And to remind each other of why you chose to work together. So whenever possible, meet with your client in person, especially if you have a long list of feedback you disagree with. When that's not possible, get them on the phone, or on Zoom. Hear each others' voices, see each others' faces. The last thing you want to be is another in a long list of emails. Everyone hates email.

Start by thanking your client for the feedback. Go over the no-brainer list. That's a quick win. Mention any particularly good suggestions they made. You need to build up some goodwill for what's coming.

With one exception: tackle the elephant in the room immediately. If one of the items on the discussion list is big enough to derail the schedule—or undoes some previously-agreed to decisions—dive right into it. Make sure your client knows the impact this issue has in terms of time and cost.

Your strategy with all the discussion items is to present your case rationally. Tell them why you disagree, use the research you did, use data, use relevant writing on the topic. Let them know how their decision might run counter to the goals of the project. And, yes, use aesthetics. That's one of the reasons they hired you, right? Make the business case, but also make the design case.

You won't win all the fights. But you might find yourself winning more of them each time you try. More importantly, you'll learn which ones are worth fighting for and which ones you can win. You'll also learn the fine art of horse trading. I once argued with a client for an hour over an issue I didn't care about (eventually letting them win!), because I really cared about the next issue coming up. At that point, they were so tired out and savoring their victory that they gave in almost immediately. Arguments are also things to be designed.

After going over feedback with your client, make sure you detail all of your decisions in writing, and have them reply with an acknowledgement so you're both on the same page as to where to go next.

WHEN CLIENTS COMP

You thought maybe I wasn't gonna address this? I just saved it for dessert.

There is nothing less helpful than getting feedback in the form of a comp (whether committed in Figma, Photoshop, PowerPoint, or Word). Nothing. I mean it. We've been in business at Mule for twenty years now, and this is the only thing we've ever fired a client over. (It happened once. The client refused to stop after being told on numerous occasions it was counter-productive, not to mention a contract violation.)

Your best bet is to nip this in the bud before it happens. Yes, put it in the contract. Now put it in your back pocket for safe keeping. As we've discussed before, your contract is your safety net and your life support. You should be able to avoid this problem by using something from your design toolbox. Namely, your amazing skills of persuasion.

Remind your client when requesting feedback that if something isn't working for them to point it out and go into as much detail as possible as to why they feel it's not working. Ask them to tie it to the goals agreed to earlier in the project. More than anything, their reasoning is critical to solving the problem. Being told to just do something a certain way—or worse, getting a comp of it—only means you have to reverse-engineer the whole thing and find out what you were trying to solve. Lost time, lost budget.

They might do it anyway. Now can we whip out the contract and beat them with it? Not so fast; your goal is always to come to an amicable solution first. There's nuance at play here.

There's a difference between comping to communicate and comping to substitute.

If your client took your comp and haphazardly started chopping it up and moving things around and even adding things from other sites, they're probably just trying to communicate with you. Yes, there's a better way to do it, for sure. But these are the acts of someone trying to get a message across. You yourselves are probably more comfortable showing solutions than describing them, right? That's what's going on here. It's not malice. It's still a pain in the ass, though; and it deserves a reminder that, while it may be the most comfortable way for them to get their point across, it's the least helpful way for you to receive it.

It's worth a sit down. Have them walk through the comp and explain to you why they made the changes they did. This can be used as an exercise on both sides. Either they felt like they couldn't get their point across verbally, or that you wouldn't understand it. A walk-through can solve both of those problems.

On the other hand, suppose the client delivers their feedback in the form of a totally different comp than the one you presented. "We were hoping for something more like this."

More often than not, this happens when there's an internal designer on the client team who's not happy about outside help being brought in. Put yourself in their shoes. You might even do the same thing. Don't.

My first question to you is whether you knew this designer existed already or not. If so, you needed to make an ally out of him earlier in the process. If not, you didn't do your due diligence. Either way you're in a spot.

You are now in a position where you're competing for a job you've already earned. That is unacceptable. Do not, under any circumstances, discuss the new comp with your client. Once you do you've validated its existence. Pull them aside. If you're working with a team, this is a job for your team leader to handle; it's account level. Remind the client that they've hired you for this job, and if they feel that you're no longer capable of fulfilling it, then a larger discussion is needed. See if the client is willing to play by the rules, as they signed up to do; if they are, great. You'll still need a strategy for dealing with that designer. I'd suggest keeping them close. If you can't come to an agreement then be ready to walk away. Never, ever, threaten to walk away from a job. You either walk away or you don't. But if you were hired to do a job and are now being asked to compete for it, then the job has changed. Your next call is to your lawyer.

LET'S END ON A HAPPY NOTE

Ninety-nine percent of your clients are good people with honest intentions. They may not give you perfect feedback every time, but their intent is good, and you share the same goal of doing kick-ass work together. As long as there's mutual respect you can get past a couple of hiccups in the process. There are always hiccups in the process.

In twenty years of business at Mule, we've dealt with a lot of client comping. Only once did we terminate a project because of it. And we gave the client multiple opportunities to stop. It was a great lesson to get out of the way early. You may never have to deal with it yourself, but if you do, you'll have the tools to handle it.

RESOURCES

Following are a few Items I hope will make you realize that we all have more to learn than we already know. Perpetual intellectual curiosity is the greatest resource a professional designer can have, and every item here will reveal its own set of resources to explore.

ETHICAL DESIGN

I mean, we have to start with *Design for the Real World* by Victor Papanek, which is the keystone of ethical design. The lessons are still good, although at this point the writing style is getting a little long in the tooth with all the defaults to he/him designer pronouns and giants-among-men stuff. But this just means that every book, no matter how important it might have been at one point, eventually needs to be rewritten. The future is fluid. Fuck the canon. There is no canon.

My good friend David Dylan Thomas wrote an amazing book called *Design for Cognitive Bias*. I borrow liberally from it, and you should too. Just give David his flowers when you do so.

Caps Lock: How Capitalism Took Hold of Graphic Design, and How to Escape from It by Ruben Pater. The title pretty much tells you everything you need to know right there on the tin, doesn't it?

You will learn more about ethics watching *The Good Place* than you learned at school. Plus, shrimps!

...and not to be self-serving, but I've heard *Ruined by Design* is a pretty good book. It's got a chapter called "Ayn Rand Is a Dick." I mean, c'mon.

LABOR

Subterranean Fire: A History of Working-Class Radicalism in the United States is about the people who brought you the forty-hour workweek, the weekend, and workplace safety. And if you're thinking "Boy, I'd love to only work forty hours a week and have weekends off!" yeah, you need to read this book.

In 1987, John Sayles made a movie about a coal miners' strike in 1920 in a small town in West Virginia. The movie is called *Matewan,* and it is very much worth your time. One of my all-time favorites. Plus, for you hipsters who enjoy privileged white men playing sad acoustic guitar songs, the movie stars a baby Bonnie "Prince" Billy as a boy preacher.

The Communication Workers of America (CWA) was one of the first big umbrella unions to embrace tech workers. Take a look at their site if you wanna see pictures of happy workers.

BETTER WORKPLACES

Vivianne Castillo and her team are doing amazing work in detoxifying the workplace at HmntyCntrd. The next time your boss makes some half-assed statement about how the company needs to do a better job at being transparent because they got caught being transparently racist? Raise your hand and tell them you know exactly who to bring in to help.

If they don't listen, the OSS' *Simple Sabotage Field Manual,* written during WW2 by the precursor to the CIA and airdropped over Axis-occupied territory, remains one of my favorite short books for fucking shit up at work. Yes, sabotage can be ethical.

Whistleblower: My Unlikely Journey to Silicon Valley and Speaking Out Against Injustice by Susan Fowler is a well-written but tough account of the kind of bullshit women deal with while attempting to do their jobs in tech.

INDEX

COLOPHON

The headlines are in James, a typeface inspired by signs carried during one of the demonstrations that led to Title VII of the Civil Rights Act. It's made by Tré Seals at Vocal, a foundry dedicated to increasing diversity in design. They do good work. Buy their type at vocaltype.co

The body copy is something else.

ABOUT THE AUTHOR

Mike Monteiro is an immigrant living with his family in San Francisco. He lives in constant fear of being evicted from their home so it can be rented out to some tech dickhead at four times the price, and of being run over by a driverless car while on his bike.

He was raised in Philadelphia, and would like to remind you that police horses are technically cops.

He's available to speak at your conference, teach your employees how to do their jobs, and write more books.

He mostly paints now and wishes the apocalypse would just fucking get here already.

He wants you to protect trans kids.

www.ingramcontent.com/pod-product-compliance
Ingram Content Group UK Ltd.
Pitfield, Milton Keynes, MK11 3LW, UK
UKHW040319300826
14791UKWH00003B/987

9 798989 358748